WH
GALAXIES
COLLIDE

WHEN GALAXIES COLLIDE

LISA HARVEY-SMITH

MELBOURNE UNIVERSITY PRESS
An imprint of Melbourne University Publishing Limited
Level 1, 715 Swanston Street, Carlton, Victoria 3053, Australia
mup-contact@unimelb.edu.au
www.mup.com.au

First published 2018
This edition published in 2020

Text design and typesetting by Megan Ellis
Cover design by Peter Long
Printed in Australia by McPherson's Printing Group

A catalogue record for this book is available from the National Library of Australia

9780522876512 (paperback)
9780522873207 (ebook)

To Shell, who supports my every dream …
and to my family, who gave me my love of the stars

CONTENTS

1

OUR FAITHFUL FRIEND

Have we lost our connection with the night sky? It's a question I often pondered when I lived in Sydney, a city of four million people. Bright lights filled our sky and only a few of the most brilliant stars shone through the orange night-time haze. When I look back at my childhood, in a small village in the English countryside, it was a very different experience.

By day, I was surrounded by fields of wheat, barley and canola. At night, a brilliant carpet of stars was laid out above me. My family was very active and we spent countless hours walking and cycling through the countryside. Watching the landscape change throughout the year, I developed a strong connection with the natural world.

I quickly learned to read the seasons. Blackberries ripened in September. Strawberries blushed in late June. In December, lone robin redbreasts hopped through the bare hedgerows. Hawthorn brambles created bursts of ripe berries for the blackbirds in January. And then the snow came. In spring, snowdrops, crocuses and daffodils peeked out from frozen ground. Wheat was harvested in August. Crop stubble was burned or ploughed into the soil in September, and the cycle repeated.

As my interest in astronomy came to light at the age of twelve, it added a universal dimension to my experiences. I learned, slowly at first, the appearance of the night sky and watched, fascinated, as the stars changed throughout the seasons. In summer, the bright trio of Vega, Deneb and Altair dominated the sky. In winter, Orion came to visit. Every night the sky circled steadily around the North Pole Star, also known as Polaris. Although I never knew him, this was the same night sky that my grandfather, also an astronomy enthusiast, had enjoyed. As I learned the constellations that he had taught my dad, it felt as if the night sky was something bigger than ourselves—something that connects us to our ancestors.

This connection to nature and to the stars helped me find my place in the world. When I was fourteen, it also helped me navigate one of the most memorable adventures of my young life. A group of friends from my athletics club and I were given a challenge: to navigate more than 60 kilometres by bike with no outside assistance. No maps, no technology, just our eyes, our ears and our surroundings. The plan was that we'd be blindfolded, driven out into the country and dumped, and then use only our wiles and our bikes to make it home before sunset. Looking back, I'm amazed that our parents gave permission for this adventure to go ahead at all.

One friend's dad, Joe, thought it would be the making of us. He hired a huge trailer and threw our bikes into the back. We dozen or so teenagers were sat down and blindfolded in the back seats of a convoy of cars. As we pulled out of town, Joe made a point of going round a roundabout a few times to disorient us. It was really exciting to be headed on an adventure, not knowing where we were going to end up. We must have driven for over an hour, my friends and I chatting excitedly at first, but then quickly getting quieter and more contemplative as we got deeper into the countryside, denied our usual dominant sense of vision.

After what seemed like hours, the car came to a stop and the door opened. 'Blindfolds off!' shouted Joe. The sunlight briefly dazzled us and the bikes clattered as they were unloaded. Car doors were slammed and Joe didn't hang around. He shouted, 'Good luck!' and the convoy drove off, leaving a small cloud of dust rising up from the country lane. And there we were, twelve teenage friends, in a field 60 kilometres from home.

Where do you start when you don't have a clue where you are? We needed to establish a baseline. We scoured the countryside looking for clues, but none of us recognised anything. Just a country lane, fields of ripening wheat and a distant church spire. A farmhand was fixing a fence a few hundred metres away. I sauntered up to him with youthful confidence and said, 'Scuse me, we're a bit lost. Where are we, please?' He smiled slightly, amused at this seemingly clueless bunch of kids on bikes, and pointed to the church spire about 3 kilometres away. 'That's Royston,' he said.

Royston. As a map freak, I knew it was on the western side of the county of Essex, where we lived. That told me we must have been about 60 kilometres north-west of home. So to get home, the only thing to do was to head south-east. Having no compass with us, nor any electronic gadgetry (this was 1994 and mobile phones were some way off), the only way to navigate was by the sun. My friends were town kids, with no particular interest in the cycles of the sun (or at least not one that they dared to express). I was the village kid who didn't go to school—I taught myself at home from the age of eleven—and spent my days climbing trees, wandering the countryside and looking at the stars. It was immediately clear that I needed to step up.

I knew that the sun rose in the east and set in the west, at least in late March (the spring equinox) and late September (the autumn equinox). In summer, which we were enjoying at the time, the sun rises slightly north of east in the UK. I was confident of this because I used to track the position of sunset throughout the year from the upstairs windows of my house. In the middle of winter I could see the sunset from the front window against the background of poplar trees on the horizon towards Great Bardfield in the south-west.

In summer, the sun set out of the small upstairs window to the side of the house towards Finchingfield in the north-west.

It was around 9 a.m. The sun moved towards the south around midday, so I figured that it had probably swung to approximately east. 'We'll get home if we follow the sun and bear right a bit!' I declared to my friends. Since nobody had any better ideas, they took my advice and off we went.

For the rest of the morning we happily zigzagged across the winding English country lanes, up and down the undulating valleys, riding towards the sun as it made its way south. It felt like an idyllic endeavour. We were undistracted, connected to our surroundings and navigating by the star that had created us. We felt free. Empowered by a basic knowledge of the sky, a bunch of kids had a fantastic adventure and made it home safely with plenty of time to spare.

Can you remember the very first time you looked up at the night sky?

For me, it was March 1986. I was six years old and Halley's comet was paying one of its regular visits to the inner solar system. The media were hyped for the appearance of this famous lump of rock and ice, which I knew appears every seventy-six years as a fuzzy 'star' gleaming in the sky, adorned with a brilliant tail. Halley's comet has been known about since at least 240 BCE, when it was first recorded in writing by Chinese and Babylonian astronomers. It even features on the Bayeux tapestry as a portent of the Battle of Hastings, streaking across the sky in 1066 as the Normans rudely invaded England.

My dad and I ventured out into the darkness of our front garden in a small country village, 50 kilometres from the bright lights of London, and craned our necks skyward. A hedgehog scuttled by. The lights scattered down our small street threw an orange glow onto the low clouds that broke up the sky, but brilliant piercing eyes of starlight tore their way through from time to time. My dad pointed up to one of these clearings and said to me, 'Lisa, we can't see Halley's comet tonight—it's just too faint—but remember that we came out tonight and looked for it.' The night we didn't see Halley's comet was my very first taste of astronomy, and it felt like a special moment. The buzz around this astronomical event made it seem like people all around the world were welcoming a rare and glamorous visitor—and I was hooked.

What was your Halley's comet moment? Your first glimpse of a dark starry sky? Seeing a planet through a telescope? Most people have been fortunate enough to enjoy these moments. When we do, we feel something resembling a primal sense of wonder. Perhaps it's because the night sky is literally in our DNA. After all, every atom of carbon, nitrogen and oxygen in our bodies was created by the nuclear fusion of lighter elements in the centre of a star. Biologists might argue the phrasing of this point, but it is true to say that from a physical and chemical point of view we are literally descended from the stars—they are our parents and our creators.

Not only that, but the stars have had a major influence on human society. Every star in our night sky has contributed to our ancestors' belief systems and, later, to our astronomical science, which has helped us build an understanding of our place in the universe. The constellations we see are virtually indistinguishable from those our great-great-great-great-grandparents saw. The stars appear so static, in fact, that their patterns have been used to pass down stories, maps, myths and legends through tens of thousands of years of human history. They are a living link to our ancestors, but that's not to say they are timeless. The universe is a dynamic place. Objects are constantly moving and shifting in space. Stars are born and they die. Galaxies collide. All of this impacts on our human experience of the stars.

How many stars do you think we can see with the naked eye? Thousands? Millions even?

Only 9000 stars are visible to the human eye. Because we can see only half of these stars from any given point on Earth's surface, that means there are only about 4500 individual stars we can stare up at and see at any one time.

If, like me, you live in a city, you're probably able to see only a few dozen of the brightest stars. That's only a minute fraction of the total number of 10^{22} (one followed by twenty-two zeros) stars estimated to exist in the universe. For our watered-down city view of the night sky we have light pollution to thank, which sees an estimated 35 per cent of all energy used in street lighting going to waste by poorly directed light shining up into the sky. According to the International Dark-Sky Association, this wastage comes at a financial cost of $3 billion to the United States alone, not to mention the detrimental effects on the sleep patterns of birds and animals and on our enjoyment of the night sky. But don't get me started.

Assuming we are in a dark place such as a mountain-top away from city lights, why can we *still* only see 9000 of the 10,000,000,000,000,000,000,000 stars with our eyes?

Most of them are too far away. The stars that we see night after night live mostly in our local neighbourhood, within 1000 light years[1] (10 million billion kilometres) of Earth. The distant stars are faint because stars give off light in all directions. That means the starlight is watered down through the colossal volume of space between the stars and Earth, rendering stars that are more distant than about 1000 light years too faint to make out. There is a mathematical rule that governs this dilution effect called the *inverse square law*. It tells us that as the distance to a star increases, the brightness fades by the square of the increase in distance. That rather dry description means that if you look at a star ten times further away, it will seem 100 (ten squared) times fainter. Put a star 1000 times further away and it has 1000 squared, or a millionth, of the brightness. That's why most of the stars in the universe are invisible to the naked eye.

---

1 A light year is the distance travelled by a beam of light in a year, or 9.5 trillion kilometres. Since the kilometre is far too small to describe the vast distances of interstellar space, the light year is one of the most common units of distance measurement we use. It also sounds pretty cool, doesn't it?

To reveal more distant stars, modern astronomers use a telescope with a large mirror or lens to collect extra light. This increases the size of our virtual 'eye', revealing far fainter objects. Without telescopes, we are left exploring only the nearby avenues of our galactic city, as our ancestors did for millennia—observing our few thousand bright stars and trying to make sense of their patterns with stories, tales and legends.

For centuries, the night sky has been divided into constellations, or groups of stars. Like join-the-dot puzzles, constellations represent the imagined shapes of mythological creatures or animals that had some meaning to our distant ancestors. They were dreamed up in ancient times and were connected to complex stories or legends. Constellations are not physical groups of stars, nor are they linked by gravity. Much like a psychiatrist's ink-blot pattern, the shapes each culture saw probably had more to do with the community's psyche than anything else.

One of the most famous constellations is Orion, the hunter, who straddles the celestial equator and is therefore visible from every part of Earth. Orion has a stout, rectangular body made up of the supergiant stars Rigel, Betelgeuse and Saiph as well as the female warrior star Bellatrix, which is now known to be a close pair of bright stars. Orion's right arm holds up a shield, and around his waist he has a belt made up of three bright stars. Below this belt is a sword made up of another triplet of stars. If you look very carefully, you can see that the central 'star' of Orion's sword is actually a cloudy smudge of light. With a pair of binoculars or a small telescope you can make out that this smudge is a cloud of interstellar gas with a cluster of bright blue stars at its core. We now know that this is a gigantic complex of 4 million trillion trillion tonnes of star-forming gas, including the famous Horsehead nebula, which is giving rise to new stars by the gravitational collapse of the cloud.

I spent many hours as a teenager peering through a 10-inch telescope and sketching the Orion nebula in the freezing cold of my back garden. Now, as a professional astronomer, I have studied its chemical make-up and its ongoing star-forming processes using some of the most powerful radio telescopes in the world.

For centuries, ancient cultures have passed down knowledge using patterns in the stars. We know little about these stories because few physical records remain and many oral stories have been destroyed by colonial practices. However, there are many strands of information remaining in cave paintings, bone carvings and oral stories from far beyond the written records.

The ancient cave paintings of Lascaux in France are a remarkable glimpse into the relationship of Neolithic humans 15,000 years ago with the stars. The paintings include depictions of several constellations, alongside pictorial representations of the creatures with which they were associated. One of these shows Taurus, the bull, with the bright red supergiant star Aldebaran as its eye. Another painting depicts a shaman (a half-human horned creature) made up of the modern constellations of Gemini and Orion.

The Wiradjuri people of Australia had another name for the constellation the Greeks called Orion. He was Baiame, a creator god who held aloft a boomerang and a shield. Orion is upside down as seen from Australia, and the Wiradjuri story is that Baiame tripped and fell over the horizon, which explains his strange orientation in the sky. Other Aboriginal nations knew him as Nyeeruna, a hunter who spent his time chasing the Seven Sisters (the star group the Greeks knew as the Pleiades).

Today, eighty-eight constellations are officially recognised by the International Astronomical Union (IAU), the worldwide professional body for astronomers that has the sole discretion to name astronomical objects. The majority of constellations (those in the northern sky that are visible from the Mediterranean) are attributed to the ancient Greeks. This is a bit of a distortion of history, though, because many of the constellation names, as well as the common names of many bright stars, are actually derived from Mesopotamia (modern-day Iraq, Turkey, Iran and Syria) between 4000 and 5000 years ago.

The Arabic names of individual stars have been honoured by the IAU, but the Greeks had a near monopoly on the official names for the constellations. Breaking this cartel in star naming took a long time, but in 2017 another eighty-six star names from Australian Indigenous, Chinese, Coptic, Hindu, Mayan, Polynesian and South African cultures were given the official stamp of approval. Finally, the academic world is starting to recognise the historical contributions of a wider variety of human cultures in the study of the universe.

The role of Middle Eastern astronomers was not restricted to inventing star patterns—it also included studying the motions of the sun and moon, predicting eclipses, and charting the cycle of the planet Venus, which, just like our moon, goes from full to half to crescent before disappearing between Earth and the sun. They also developed the method of dividing the virtual 'sphere' of Earth's sky into 360 degrees—the basis for much of our modern mathematical way of measuring the sky.

Civilisations across the globe each have their own interpretations of the star patterns. Ones you may not have heard of include the llama (Inca), the ear of grain (Persian) and the white tiger of the west (an ancient Chinese asterism). The ancient Chinese developed their own sophisticated set of constellations some 2500 years ago, which in subsequent centuries (after contact with Middle Eastern astronomers) was adapted to incorporate some of the star groups from the Mesopotamian tradition. The Chinese have arguably kept the most detailed scientific records, including noteworthy records of the stars and planets. They are the only civilisation to have recorded every appearance of Halley's comet from 3000 years ago to the present day.

The ancient constellations were not just the product of artistic minds gazing at the stars and imagining famous people and animals. They were also connected to spiritual beliefs. Many of the early stargazers were paid by kings, queens and emperors to help them make important decisions. In those days, running a kingdom based on astrology—the alignment of stars and planets—was a popular way to conduct political affairs. Some fascinating social rituals also arose from the very deep belief many cultures had in astrology. That the word 'disaster' has its roots in the Greek for 'bad star' points to the importance of astrology in human history. In Mesopotamia and Iran, people employed trickery to counter the impending arrival of 'unlucky' astronomical events such as eclipses. To bamboozle the evil forces, the astrologer in the royal court would arrange for a powerless subject such as a peasant, a prisoner or a political opponent to sit on the throne for the duration of the 'unlucky' event. The idea was that the bad luck would fall upon the fake king, while the real monarch, tucked up in the castle or palace, would be spared. Sadly, the event always turned out to be unlucky for someone, because the substitute monarch would end up being sacrificed once the supposed danger had passed.

Of the eighty-eight official constellations, those in the Southern Hemisphere were named most recently, often by travelling Europeans. Dutch-Flemish cartographer Petrus Plancius invented twelve new constellations based on maps of the southern stars drawn up by Pieter Dirkszoon Keyser and Frederick de Houtman on their travels to the eastern Indian Ocean in 1595. These new constellations included exotic creatures such as Apus (the bird of paradise), Chamaeleon (the chameleon), Dorado (the dolphinfish), Pavo (the peacock), Phoenix (the mythical phoenix) and Tucana (the toucan). They left us with a veritable menagerie in the southern skies.

French astronomer Nicolas-Louis de Lacaille created fourteen new constellations from his observatory near Table Mountain in South Africa in 1750. Unhindered by the tradition of naming groups of stars after Greek mythological creatures or animals and birds, he decided on some more modern, technological names. Lacaille's constellations included Antlia (the air pump), Fornax Chemica (the chemical furnace), Microscopium (the microscope), Telescopium (the telescope) and Horologium (the clock).

The most famous constellation in the Southern Hemisphere today is the Southern Cross, which currently appears on the national flags of Australia, New Zealand, Samoa and Papua New Guinea. The Southern Cross is the smallest of the constellations but is easily recognisable because its stars are all so bright. It has been well known by people of the Southern Hemisphere for thousands of years; they used it for navigation. Maori and Australian Aboriginal peoples had various names and stories for the constellation, calling it the anchor, the stingray or the great sky canoe. Ancient Greeks also knew these stars, but they only just grazed the horizon from southern Europe and Greek astronomers lumped them in with the neighbouring Centaurus constellation. They were not given the name 'Crux' (Latin for cross) until the 1500s, when their full glory was seen by travelling European traders.

One of the most notable Southern Hemisphere stars is Alpha Centauri, which is actually made up of two stars that orbit so close to one another (about the distance between Earth and Neptune, on average) that we can't distinguish them without a telescope. A third star, called Proxima Centauri (from the Latin *proximus*, which means 'nearest') is loosely gravitationally bound to the pair and is currently the closest star to Earth at a distance of 4.3 light years.

The constellation of Ursa Major, the Great Bear, is my personal favourite. It goes by a few names, most commonly the plough or the saucepan. The plough was always visible out of my bedroom window when I was growing up, and I used to watch it spin around the North Pole throughout the night. It is connected to the Pole Star by Ursa Minor, the Little Bear, which is a smaller constellation that looks (at least to me) like a giraffe. Had the ancient Greeks ever seen a giraffe, I have no doubt it would have that name. Since I've moved to Australia I miss my old friend Ursa Major—it is too far north to be visible in the south-east corner of Australia.

Ursa Major was also the first constellation taught to me by my dad one cold night in Essex. We soon learned how to use it to navigate—following the two right-hand stars of the saucepan directly upwards to find the North Pole Star, Polaris. My dad learned its distinctive shape from his own father on their night-time walks in

Wolston, the village where he grew up. My grandfather died shortly after I was born, but I'm happy that we are able to share our love of the night sky in this way.

Do we still see the night sky the same way that the ancient astronomers saw their constellations over 2000 years ago?

Perhaps surprisingly, the answer is yes. Most of the Greek constellations appear very similar today to when their ancient legends were first written down. There are some small differences, however. So what changes do we see in the stars throughout history, and what elements of the night sky stay the same?

Astronomers of the ancient Greek and Roman empires believed that the stars were attached to an invisible dome called the firmament, a sort of solid crystal sphere that encircled Earth. Many of the classical Greek astronomers believed the dome was made of a transparent material called 'quintessence' or 'aether', a supposed fifth element of nature (the others being earth, air, fire and water) that existed in the universal realm to keep up the sky. Their theory, which had no real evidence behind it but seemed to explain the motion of the heavenly bodies, lasted well into the seventeenth century. The constellations, explained Greco-Roman astronomer Ptolemy, remained fixed in their positions as the firmament rotated above our heads.

This model was later adapted to include the planets. If you look at one of the bright planets such as Mars, Jupiter or Saturn today, you'll notice that it moves very slightly each night relative to the constellations. For example, from my view of the sky tonight, Saturn is in the constellation of Ophiuchus. Next month it will move into Sagittarius. That is because Earth and Saturn are both moving through space as they orbit the sun, so the relative positions of our planets in the sky changes slightly from night to night.

The stars are millions of times further away than the planets, so their motions appear absolutely tiny in comparison. Essentially, the stars appear 'fixed' to the ceiling whereas the planets earned the name 'wandering stars'. They saunter around the sky throughout the year as they (and we) move around the sun, and slowly vary their brightness as they move closer to and further from Earth.

Not knowing about the orbital motions of the solar system, the ancient Greeks thought that each of the planets was attached to a fixed crystal sphere or 'orb' surrounding Earth. In this complex scheme, the sky was imagined as an onion-layered nest of crystal orbs, each carrying its own planet, with the firmament on the very outer layer carrying the fixed stars. This theory got pretty contrived.

Around 2000 years ago, astronomers began to notice drift and changes in the positions of the stars. Through careful observation and measurement, we came to know that the stars are free to move about throughout space, unencumbered by the crystal domes that the Greeks imagined. We now accept that Earth itself has no special or 'central' location around which orbs of stars and planets dance. The system of the firmament, fixed stars and orbs was quickly dumped. Space had become officially complicated.

Our night sky is changing, albeit at a glacial rate through the ages. The reason for the changes is a natural effect called the *precession of the equator* or, more commonly, the *precession of the equinoxes,* figured out by the meticulous work of a Greek astronomer called Hipparchus in 127 BCE. He studied the position of the sun and stars using the 'precision instruments' of the time, including the gnomon (essentially a sundial) and a set of spheres and circles representing the paths of the sun and stars in the sky called an astrolabe or armillary sphere.

Comparing his measurements with those made by his predecessors Timocharis and Aristyllus around 300 BCE, he noticed that the whole sky had moved by an angle of about 2 degrees. The orientation of the sky, he surmised, was systematically moving year on year. And he was right—this drift is still seen today. So how does this happen?

Earth spins once per day on its axis, an imaginary line that passes through the North and South Poles. Earth's equatorial region spins at about 1670 kilometres per hour, as opposed to the more sedate

1260 kilometres per hour at latitudes of around 42 degrees north and south equivalent to New York City, USA or Hobart, Australia. At higher latitudes still, such as 60 degrees (equivalent to the Norwegian capital, Oslo), the rotation speed drops to 837 kilometres per hour.

This is why rocket launching sites are often located close to the equator—the faster rotation of Earth's surface at these latitudes gives the rocket an extra 'kick' to get into orbit. Also as a result of this speed differential, Earth is slightly bigger at the equator (12,756 kilometres in diameter) than at the poles (12,713 kilometres in diameter), because as it rotates it fattens slightly in the middle.

The sun and moon exert a small gravitational pull on Earth's belly paunch that creates a turning force, or torque, on Earth that causes the axis of Earth to slowly draw out an invisible circle in the sky once every 28,500 years. This motion can now be measured incredibly accurately using networks of radio telescopes spread across the globe.

How does precession of the equinoxes play out in reality?

Over thousands of years the north and south celestial poles wander very slightly year on year. Today, if you watched the Northern Hemisphere sky for twenty-four hours, you would see that all the stars appear to rotate once around the bright star Polaris in the constellation of Ursa Minor. That's because at the moment the North Pole of Earth happens to point towards Polaris, so every rotation of Earth on its axis appears to occur around that star.

Back in 12,000 BCE, the North Pole Star was not Polaris at all. Due to the precession of Earth's axis through the sky, it was closer to the bright star Vega in the constellation of Lyra (the harp). As Earth's axial precession continues, Vega will again take up the mantle of North Pole Star in about another 13,500 years. Around the same time, the second-brightest star in the sky, Canopus, will come within 10 degrees of the south celestial pole. Bright stars like Vega and Deneb, which are currently seen from Australia in the winter, will be too far north to be seen from these shores. Also in that era, the constellation of Orion will have shifted so far south that it will be invisible from northern Europe, northern Asia and parts of the USA and Canada. All in all, precession will have a significant effect on how our night skies look from any given place. It will not affect the shape or make-up of the constellations, however—*only our viewing angle on the stars.*

But there are other forces in play that change the night sky over the generations—silent drifts and dramatic events that can alter the constellations as we know them.

Edmund Halley was an English explorer, astronomer and scientific all-rounder who was interested in mapping the stars. In 1676, as a twenty-year-old undergraduate student, he came up with the idea of mapping the southern stars. King Charles II was so keen on this idea he ordered that Halley be given free passage on a ship called *Unity* to the volcanic tropical island of Saint Helena in the South Atlantic. Halley set off without finishing his degree, but this didn't seem to bother him. He spent more than a year using a naval navigation instrument called a sextant to measure the positions of stars with respect to each other and the horizon, producing one of the English-speaking world's first surviving maps of the stars in the Southern Hemisphere. His map provided an important and accurate account of the southern stars in the late seventeenth century. He even named a new constellation Charles' oak after the king. The gesture seems to have been well received, as Charles II ordered Oxford University to award Halley's unfinished degree honorarily.

Later, Halley studied the historical records of all the comet sightings from the past 350 years. Looking at the motions of historical comets around their closest approach to the sun, he noticed the comets that had appeared in 1531, 1607 and 1682 CE had very similar orbital paths. His stroke of genius was to realise that these were probably all the same comet, which returned to its closest point to Earth as it orbited the sun in an elliptical path every seventy-four to seventy-nine years.

At the time, this was a great leap of imagination, since comets had always been assumed to appear once and once only. Halley's theory was vindicated in 1758, sixteen years after his death, when the comet turned up on Christmas Eve. It quickly became known as Halley's comet, named in honour of the man who had first realised that comets can have stable orbits. It's just a pity that Halley himself didn't live to see it for a second time.

Another of Halley's triumphs was detecting anomalous motions of one or two bright stars through space. In 1718 he announced to the world in a paper in the academic journal *Philosophical Transactions of the Royal Society* that these two 'fixed stars' actually had a small but measurable motion with respect to the others. This was a controversial claim at the time because it broke down the commonly held belief that the heavens were fixed and perfect. The facts were indisputable, however, and the newly discovered effect was called 'proper motion'. It was Halley's painstaking measurements and historical detective work that enabled him to notice shifts in the stars' positions.

When he compared his star maps to those made almost 2000 years before by the Greek astronomers Ptolemy and Hipparchus, Halley noticed that in addition to all the stars systematically moving due to the precession of the equinoxes, two of the brightest stars, Sirius and Arcturus, had also independently moved out of whack with the other stars by approximately the width of the full moon.

Next time you look at the moon, check out its width—it's smaller than your little fingernail at arms length. I find it pretty astonishing that Halley was able to notice such a small shift in the position of these two stars. Over thousands of years some stars will shift their positions quite noticeably, eventually stretching and squashing the shapes of the constellations.

With modern scientific equipment it is far easier to measure the relative motions of stars, and on a much larger scale. The Hipparcos satellite, launched in 1989, measured the positions and brightnesses of 100,000 stars 100 times over a period of four years to track any changes. This ultra-accurate process gave us an insight into just how much the constellations are changing and will change over the next 100 millennia.

Some constellations, such as Orion, will remain pretty much the same over the coming 100,000 years, with the exception of one or two of the bright stars moving out of place. Other constellations, like the Southern Cross, will completely unfold and become unrecognisable over the course of the next 50,000 years. The reason for this variation is because most distant stars move very little through the sky but some nearby stars have larger relative motions.

Given that the relative motions of most stars are quite small, the majority of the constellations as we know them today will last tens of thousands of years before their shapes dissolve. It's pretty amazing to think that you and Julius Caesar could share such a similar view when up on a hillside gazing at the stars!

Despite the stars' reputation as steady and unchanging, astronomers through the ages occasionally recorded 'new' or 'visiting' stars in the night sky. The nova and its more dramatic cousin, the supernova, are the modern names for these elusive visitors to our skies.

In his book *Kafka on the Shore*, Haruki Murakami described the stars as alive and breathing, like trees in a forest. A poetic phrase, but could there be any scientific truth in it?

There is no doubt that stars have a life. Like us, they are born and eventually die. Their lives begin inside huge clouds of interstellar gas that collapse into spherical clumps under the influence of gravity and eventually achieve the critical mass that triggers nuclear fusion processes that make the stars shine for hundreds of millions or even billions of years. Many shine steadily, but others pulsate wildly, breathing in and out on a regular schedule like a giant pair of lungs. Once the fuel that sustains the nuclear fusion is exhausted, stars begin to die. They either fizzle out as white dwarfs in the case of smaller stars like our sun, or explode catastrophically in supernova explosions in the case of very massive stars.

Every year or two on average an apparently 'new' star will appear in the night sky. It will brighten suddenly and without warning before slowly fading from view over a period of a few weeks. An event like this is called a *stellar nova* (new star).

Most stellar novae are quite faint. To the casual observer, the appearance of another average star within a cast of thousands is barely noticed. But every now and then a really bright nova appears—something to rival the brightest stars in the sky. These are really special events, happening only a handful of times per century. And they certainly do not go unnoticed. They have been recorded as long as people have been looking at the stars, and appear in oral stories, ancient petroglyphs, and writings from the ancient astronomers. They are a part of our astronomical history.

Novae are not 'new' stars at all. We now know that they are caused by a flare-up in the brightness of a star as it interacts with a stellar sibling. As many as 70 per cent of all stars live in pairs or multiple-star systems, where two or more stars live in a stable orbit around their shared centre of gravity—caught in a cosmic twirl, if you will. A stellar nova occurs when some of the material from one of these stars is dragged by gravity onto the surface of the other, leading to a sudden increase in the fuel load available to the recipient. This causes a huge and sudden increase in the nuclear fusion process happening in the star, and its luminosity grows 50,000-fold. Once the extra fuel is used up, the star slowly returns to normal. To reflect our modern understanding of stellar novae, they are now called *cataclysmic variable stars*.

Cataclysmic variables are only one of a whole host of variable objects in our night sky. These include colliding stars, exploding stars, stars feeding black holes, merging neutron stars and black holes that generate gravitational waves, and other objects whose nature is as yet unknown. Studying them has become a huge industry. When a new bright object is discovered in the night sky, astronomers quickly release an 'astronomical telegram' (actually an email), sending the details of the object (position, brightness) instantaneously around the world. This prompts us to immediately mobilise our telescopes, sometimes within minutes, to track the object as it brightens and fades. You might call us cosmic storm chasers! By using a variety of instruments working in optical light, X-rays, radio waves and other flavours of radiation, we can build up a clear picture of the behaviour of variable astronomical objects as they undergo their extraordinary transformations.

In 2011 I was observing some interesting galaxies using the Australia Telescope Compact Array near Narrabri in New South Wales. On the second night of my week-long stint I received an email from a fellow astronomer, Stéphane Corbel, from the Service d'Astrophysique in Paris. He had received an astronomical telegram alerting him to the appearance of an outburst of radiation from a pair of black holes in our galaxy that orbit one another. He wanted to quickly monitor the radio emission that was being given off by the event. Although my time on the telescope had been twelve months in the planning, the opportunity was too good to miss. I duly obliged, giving up my galaxy studies for the evening and turning the telescopes to study this remarkable event. The result was an excellent scientific study that followed the ignition of jets of material streaming from the region around the black holes as they interacted. Although it was not quite what I had planned, the cosmos is master in these matters!

A stellar nova might sound bright, but it can easily be upstaged by its more famous and dramatic cousin, the supernova. A supernova marks the almost total destruction of a star by a cataclysmic explosion. It is an extremely rare event, lasting only a few weeks and happening to fewer than 1 per cent of the most massive stars.

So what causes a star that has shone steadily for thirty million years to suddenly and dramatically explode? And could it happen to our sun?

Supernovae can occur in two different ways. The first, called a type Ia (one-a) supernova, happens in a similar way to a stellar nova. When a white dwarf star (already at the end of its life) sucks material from an orbiting star onto its surface, it can flare up as a cataclysmic variable or, given the right conditions, can start off a chain reaction that leads to complete and total destruction of the star. If enough material has fallen onto the white dwarf to raise its mass above 1.44 times the mass of the sun, the core reaches a critical temperature that allows carbon nuclear fusion reactions to begin in the core. This sparks an enormous increase in temperature.

Since white dwarf stars are made of neutrons, they are unable to expand and regulate their temperature like normal stars. Consequently, the heat causes a shock wave that travels at about 10,000 kilometres per second and blows the star apart. It shines for several weeks, peaking at around five billion times brighter than our sun.

The other type of supernova is the core collapse or type II (two) supernova. When a star has more than about ten times the mass of our sun, it burns 3000 times more brightly and lives only thirty million years, in contrast to the more pedestrian pace of our sun's hydrogen-to-helium nuclear fusion, lasting around nine billion years.

As the hydrogen fuel runs out, it turns to helium, then to carbon, neon and oxygen, then finally to silicon as a source of fuel. Each step produces greater and greater heat in the core of the star. The final silicon fusion reactions burn furiously at a temperature of 3 billion degrees Celsius for only one day, in which time nickel-56 is formed in a process called silicon burning. The nickel quickly undergoes radioactive decay into iron. Due to the chemical nature of iron, it takes more energy to burn than it releases—in other words, it is a hopeless fuel. At this point there is literally nothing left to burn and the core of the star collapses under its own weight at a velocity of 70,000 kilometres per second. Its temperature hits 5 billion degrees Celsius and it becomes so squashed together that all the electrons and protons combine to form neutrons (the neutrally charged particles in the nuclei of atoms). Finally, the core becomes so dense (equivalent to the density of an atomic nucleus) that a very powerful force called neutron degeneracy pressure kicks in, slamming on the brakes. The enormous energy of the core collapse is halted and the incoming material rebounds outwards. The star's material is expelled at about 30,000 kilometres per second, and the star goes supernova. It shines with a luminosity ten billion times brighter than the sun. During this process, elements heavier than iron are formed. All heavy elements on Earth, such as gold and silver, were formed either in supernovae or in other cataclysmic stellar explosions.

Fortunately for us, neither of these scenarios will play out for our sun. But supernovae do happen from time to time in our neck of the woods. We have historical records of naked-eye supernovae, happening on average every 200 years. There are stories in rock art and oral traditions about the appearance of 'new stars' throughout human history. Chinese records scratched onto a bone from 14,000 BCE talk about a brilliant 'new star' close to Antares. A rock painting in New Mexico created by the indigenous Anasazi people of North America shows a crescent moon above a brilliant multi-pointed star. Calculations show that the supernova recorded by Chinese and Japanese astronomers in 1054 CE would have first appeared below the crescent moon at the Anasazi site, but many historians agree that the star might actually represent Venus.

The first records definitively linked to a supernova were for an event in 185 CE. Supernova 185 was described in the Chinese text *Hou Hanshu* as a guest star that appeared close to the bright star Alpha Centauri on 7 December 185 CE. It shone as large as half a yan (a coin) in scintillating multicolours. The new star shone brightly for a period of eight to twenty months (the historical record is unclear) before finally shrinking and fading from sight. Many astronomers link this type Ia supernova with the supernova remnant called RCW 86, a glowing relic of a stellar explosion seen in the same region of the sky that has expanded to 85 light years in diameter.

After SN 185 not much happened on the supernova front for another 200 years. New stars were recorded in 369, 386 and 393 CE by Chinese astronomers, according to the *Song Shu* historical text. Unfortunately, little detail about the colour, size or brightness of these stars made it into the records. The duration of the 393 CE event was reportedly eight months, suggesting that it was a supernova, while the events in 369 and 386 CE lasted for somewhat shorter periods and it is possible that either or both of them were just stellar novae.

The next dramatic supernova was seen on 30 April 1006 CE. It shone incredibly brightly for several years and dominated the night sky. The scholars of the time wrote that it was so bright it was even visible during the daytime. Chinese records describe it as 'huge … like a golden disc' and 'like the half-moon … it had pointed rays'. One account said that you could read a book by its light. The star was so brilliant it appeared in records throughout the world, from Japan and China to the Middle East, Africa and parts of Europe. It was one of the most extraordinary sights in the night sky ever witnessed by humans.

In 2007, 1001 years after that explosion, I studied the region of the sky where the great supernova of 1006 CE had been seen. Using a group of six 30-metre-diameter radio telescopes in Narrabri, I made a detailed image of the remnant of the 1006 CE supernova. My image showed a huge bubble of gas, about the size of the full moon, with a clear, sharp boundary marking the shock wave of the supernova that has been expanding into space for the past 1000 years. As I held the image in my hands, it was hard not to feel an overwhelming sense of history—my own part of the astronomical record.

Further bright naked-eye supernovae appeared in the years 1054, 1181, 1572 and, finally, 1604 CE. Today, we seem to be long overdue for a supernova. Let's hope it's a spectacular one and bright enough to impress. But let's not wish for anything *too* spectacular. The colossal energy of a nearby supernova would release dangerous high-energy particles, such as cosmic rays and gamma rays, that would stream towards Earth at the speed of light. Gamma rays have the potential to disrupt the ozone layer and subject Earth to elevated levels of high-energy radiation. This in turn could wipe out several species, including phytoplankton, which have a crucial role in oxygen production on Earth and form the basis of the marine ecosystem. Such a large disruption to the oceans could have a devastating knock-on effect on life on Earth.

Apart from the occasional nova or supernova, some stars vary their brightness more gradually over time. More than 50,000 of these *variable stars* are now known. Due to the efficiency of modern photographic techniques, their detection has become no more than an automated process. However, in pre-photographic times it took a far more systematic approach to astronomy to detect changes in the brightness of stars.

The first[2] recorded sighting of a variable star was in 1596 CE by David Fabricius, an amateur astronomer from Friesland in the Netherlands. He was looking for the planet Mercury through his backyard telescope when he picked out a dull star in the constellation of the whale (Cetus) to carry out a brightness comparison. Two months later, when looking for Mercury again, he found that his randomly selected comparison star had gone. More than a decade later, while repeating the same exercise, he found that it had reappeared.

This behaviour was not understood until forty-two years later, when Johannes Holwarda, another Dutch astronomer, located the star once again in just the place that Fabricius had recorded it. He began measuring its brightness over a period of weeks and months—again, by comparing it to other nearby stars of known brightness. He realised, after studying the star for a couple of years, that its brightness varied within a regular period of eleven months.

The name of this star—Mira, meaning 'wonderful' or 'astonishing' in Latin—was coined by a beer-brewing Polish-German astronomer called Jan Heweliusz, who also monitored its slow blinks. Mira's

2 The variable star Mira may have been discovered as early as 100–300 BCE, but original star maps and brightness records from that time are missing or incomplete, so this is uncertain.

behaviour means that it certainly deserves such a title, and the name stuck.

So why does Mira vary so dramatically?

Well, we now know it to be an old, giant, puffed-up, pulsating variable star that slowly sheds its outer layers of gas into deep space. As it expands, it cools and becomes fainter before contracting and heating once again in cyclical fashion. Mira is truly a living, breathing, pulsating entity.

Another type of star that appears to vary its brightness is an *eclipsing binary star*. Its poster child is Algol, in the constellation of Perseus. Algol dims and brightens every two days, twenty hours and forty-nine minutes, with clockwork regularity. In ancient times, it was called the Demon Star and considered to be very 'unlucky' (whatever that meant). It wasn't until it was studied using a large telescope that astronomers realised it was a pair of stars orbiting their common centre of gravity. As the fainter star in the pair passed in front of the brighter, the total brightness noticeably dimmed. This type of system is called an 'eclipsing binary' and works in the same way as a solar eclipse, where the moon passes in front of the disc of the sun. See? With scientific knowledge, the demon is tamed and there is nothing to be scared of at all.

There is one feature in our night sky that remains remarkably stable and quiescent: the band of light that we call the Milky Way.

The Milky Way as a feature of the night sky has remained unchanged for millennia. Unlike the stars that make up constellations, the Milky Way's stars can't be distinguished individually with the naked eye. Instead, the combination of 100 billion stars combines to produce a veil of light, stretched out like distant city lights seen from an aeroplane at 40,000 feet.

Witnessing this magnificent silvery river stretching from horizon to horizon is a breathtaking spectacle, and it takes a very dark sky to appreciate it fully. Like the brushstroke of an artist, the Milky Way has an exquisite character. The neat band of light comes from the tightly confined plane of our galaxy. It is broken only by coal-black cloudy patches caused by vast interstellar clouds of molecular gas that block the light from the stars behind them. The sheer number of stars in the Milky Way means that variations in the brightness of a few stars do not noticeably change the appearance of the whole.

The Milky Way has its own astronomical folklore. With no telescopes at their disposal, people prior to 1610 (when Galileo Galilei first turned a telescope to the night sky) could not have known that it was made up of billions of stars. Instead, they came up with a host of colourful names and stories, many of which reflect our galaxy's appearance as a path or road through the stars. The Greek story described a path of milk, spilled by the goddess Hera while she was breastfeeding Zeus's overenthusiastic baby Heracles. The Greek *galaxias kyklos* means 'milky circle', which is the derivation of the word 'galaxy' that we now use to describe any large, gravitationally bound collection of stars. This was translated into Latin as *Via Lactea* or 'Milky Way', which gives us the common name for our galaxy in many languages, including English.

There are many other names and descriptions of the Milky Way across the globe. Nordic cultures called it 'the winter way', since at high latitudes the faint glow was only visible in winter. In Finland, it was called 'the path of the birds'. On the Arabian peninsula and in the Andes, India and East Asia, people saw it as a river. Large swathes of the Middle East, Africa and parts of the Balkans spoke of 'the route paved with straw'. The Cherokee of North America called it 'the way the dog ran away', and in Thailand it was 'the way of the white elephant'. Some of my ancestors, the Anglo-Saxons, called it 'Watling Street' after the trading route (which later became a Roman road) that crossed England from Chester in the north-west to the port of Dover.

As some cultures imagined patterns in the stars, others based their stories on the shades of light and dark within the Milky Way. For example, many Indigenous cultures on the Australian continent saw the shape of a great emu in the dark clouds of the Milky Way. Even though I'd never noticed it before, after the emu was pointed out to me it became impossible to forget. In a similar vein, the South American Incan culture built its mythological stories around creatures such as the llama and snakes that appear in the patterns of dark dust clouds against the backdrop of the Milky Way.

So how has scientific inquiry led to a sophisticated understanding of the Milky Way and our place within it, and our galaxy's turbulent future?

2

ON THE INSIDE, LOOKING OUT

People sometimes ask me, 'What do you do all day?' It's a fair question for someone who earns a living staring into space. Do I sit on a mountaintop at night and look through a telescope at the stars? Do I count them? Take pictures?

As an astrophysicist, my job is simple, really: I design and conduct experiments to figure out how the universe works. Like Isaac Newton said, if we achieve greatness (or anything useful), it is only by standing on the shoulders of giants. Drawing ideas and inspiration from reading scientific journals and sharing knowledge with colleagues at conferences, I identify interesting unsolved problems in astronomy and try to find solutions. For me, that usually means drawing on a combination of new and existing observations of the night sky using large radio telescopes.

To be granted access to a multimillion-dollar telescope, I need to write a proposal to the observatory that operates it. Proposals are accepted every six months or so and reviewed by a panel of experts, and the authors of the best ideas are then awarded a set number of hours to conduct their observations. I used to chair the committee for the Australian Commonwealth Scientific and Industrial

Research Organisation (CSIRO) that awards observing time on Australia's radio telescopes; it was a big job but a rewarding one, and it gave me a great knowledge of the research that was being carried out across my discipline.

Until very recently, when the internet age hit, we astronomers used to travel to the telescopes and steer them using control panels and specialised computer software at the observatory. It was an unusual job, requiring skill and patience as you troubleshot hardware and software problems as they inevitably arose during the day and at night. Operating the telescopes for our research involved draining shifts of up to sixteen hours and a lot of travelling to remote and exotic locations, since world-class radio telescopes are traditionally sited in remote areas such as rural hamlets and, increasingly, outback deserts. Their distance from inhabited areas is important to retain the radio-quiet conditions conducive to listening for whispers from distant stars and galaxies over the foreground cacophony of radio signals booming out from human activities on Earth. It can, however, present challenges to young and inexperienced researchers travelling alone to what can often be difficult or even hostile remote environments.

For instance, as a young postdoctoral researcher, I was sent to Puerto Rico to study the structure of the Milky Way using the 305-metre-diameter Arecibo radio telescope. It was the world's largest telescope at the time, and for a young astronomer who had grown up revering Arecibo as the world's greatest scientific instrument, it was a mouthwatering adventure. After a 24-hour journey from Australia I arrived at the Puerto Rican capital, San Juan, tired but excited about the unfamiliar environment and the prospect of using a telescope I had read about in books since I was fourteen years old.

The taxi skirted the suburbs of the capital and began to snake through the countryside towards the small town of Arecibo. After a little while, the driver took a detour into a residential area and pulled up at a house, then jumped out without so much as a word.

At that moment my heart sank. I realised that, as a young female traveller, I was completely at the mercy of this stranger and things were not going as they should. Thoughts raced through my mind, and my heart raced too, as I contemplated my complete helplessness. I started to make escape plans, albeit rubbish ones like 'I'll just run off if he tries anything'. After a couple of minutes, I saw him return alone and hop back in the driver's seat. We continued our journey without a word. I never did find out why we stopped, but I wasn't in the mood to ask questions.

Normality resumed as I sailed through the countryside in the back of the taxi, surveying the banana and yukka trees and the roadside shacks selling groceries and cold beer. After about half an hour we pulled up at the famous observatory's boom gate to be greeted by a security guard. It was a bit of a culture shock, to say the least, seeing an armed guard watching over the almost-deserted jungle observatory. He was friendly, though, and welcomed me warmly before driving me up a steep, winding track to the observer's quarters.

As the sun dipped below the domed volcanic peaks that encircle the telescope and provide natural shielding from the stray radio waves generated in the towns strewn across the island, I felt like I had arrived in paradise. Thanking the guard for the lift, I dragged my case into my quarters—a rustic wooden cabin that I recognised from the Hollywood movie *Contact,* which was filmed at the observatory. *Nice touch*, I thought. For this week, I decided, I *was* Ellie Arroway, the nerdy film's radio astronomer hero, played by Jodie Foster.

I took off my shoes and lay on the bed for a moment. *I can't believe I'm here*, I thought. Excited and restless, I jumped up to check out the porch, which had two little al fresco chairs to sit on and relax in. I tried one for size. The frogs that densely inhabit the island were united in an enthusiastic chorus of 'Croak-ee! Croak-ee!' and a warm breeze brought the scented jungle air past my face.

Suddenly, a loud *SLAM!* broke the spell. The door had slammed shut, caught by the wind. I tried the door handle. It was locked. 'Shit,' I said to no one in particular. I peered through the windows of the cabin and there was the key, sitting brazenly on the desk and staring gleefully back at me.

Barefoot and alone on a dark hillside, I considered my options. Spending the night on the porch seemed OK temperature-wise, but after hearing the buzz of jungle-sized mosquitoes, I decided it was not a good option.

Mobile phones are strictly banned at radio observatories because they mess with the equipment, so calling for help was out of the question. More to the point, who would I even call? It was clear that I needed to head back down the mountain and find someone to help me get back into the cabin. 'Croak-ee!' called the frogs, encouraging my expedition.

I felt my way down the path and towards the road I had travelled up with the security guard. The bitumen was warm and a little spiky underfoot. The warmth made me think about basking snakes.

Using the soles of my feet to sense the road, I walked back down the hill towards … I didn't know what. There were two faint outdoor lights in the distance and I decided to head for those.

'Croak-ee!'

As I descended in the dark, my predicament amused me. *What a great story it will be if I get lost in the jungle and die this way. Maybe they'll name the telescope after me!* I thought.

Focus, Lisa.

At the bottom of the hill I came upon a demountable that looked like something out of the 1970s. It had one of those spherical handles on the door, which I turned. 'Please open …' I begged, and *click*. By sheer luck it opened. Venturing inside I found a light switch that when I flicked it bathed a deserted library and some comfy seats in the depressing blue hue of a fluorescent strip-light. *This will at least be a safe place to sleep*, I thought, but to be honest it creeped me out. I knew I had to find signs of life down here.

Following a series of concrete pavers, I found my way to a neighbouring building. It was locked, but the lights were on. I tapped on the window. After a short pause the night telescope controller answered the door, looking me up and down. *Hasn't this guy ever seen an astronomer with no shoes?* I thought.

'Hi,' I said, out loud this time. 'I'm a visiting astronomer from Australia. I'm observing tomorrow night. Just locked myself out of my cabin.'

His eyes sank to my bare feet as he let out a light sigh at my strange but plausible story. The security guard was summoned and barely suppressed his amusement as he picked me up. Within ten minutes I was being driven back to my cabin. 'Don't forget your key next time,' the guard mumbled redundantly before he headed back down the warm, snaking (and possibly snaky) path to his guard post. That was day one of my first observing experience at Arecibo.

The whole trip was typical of the sorts of things a young astronomer will experience. On the third night of observing (at midnight), my colleague and I donned hard hats and joined local engineers on a trip to the focus cabin of the telescope, climbing ladders and traversing suspended walkways hanging 127 metres above the 8-hectare dish surface. As on all radio telescopes, the walkways were made of steel mesh and completely transparent. Vertigo is pretty much guaranteed during the day, but in the inky blackness of midnight it was mercifully calm. With no mention of workplace health and safety apart from 'hold on', we clambered into a one-person-sized cabin and took the steel lid off our radio receiver of choice, replacing the cover on the unwanted camera, before climbing back into the engineering lift and trundling down forty-one storeys to the jungle floor to resume observing from the control building.

Other highlights of the trip included going for a run in the countryside and being repeatedly catcalled and then chased by two guys in a ute. I've never run so fast. Even our day off was risky—we told observatory staff we planned to visit a local beach and were sternly warned by locals about deadly rips, rabid dogs and armed robbers.

With the advent of the internet, all this has changed. Nowadays, most telescopes are controlled remotely. Once my observations are scheduled I simply log on, press a few buttons and monitor how everything is going using real-time streamed information. I've taken data with several major telescopes, including the one at Arecibo, from my laptop at home, wearing my best flannel pyjamas.

Some astronomers pine for the old days, but I view this culture shift as positive. Replacing fieldwork with remotely operated systems is safer and more environmentally conscious, and can make astronomy as a career more family friendly. The skill sets of astronomers are shifting, too. The ability to conduct observations is becoming less important but we are acquiring more advanced software skills. Astronomers of the future will need to process large amounts of data in smart ways that enable them to form reliable scientific conclusions. We're now anticipating the next step: entirely automated (computer-controlled) observing for our new telescopes is about to come online.

Research was very different in the past. Before large telescopes shared between many researchers became the norm in the early twentieth century, astronomers would often build their own observatories in their gardens. This isolated way of working meant that scientists would have to share information about new discoveries by writing letters to one another and publishing their results in books or via presentations to the learned societies (for example, the Royal Society in London). Without particularly high-tech equipment, it's amazing how much was achieved, including the first study of the Milky Way's structure.

The Milky Way is 13.2 billion years old, almost as old as the universe itself. Complex and fascinating, it is just one of hundreds of billions of galaxies, a spiralling city of around 200 billion stars. We live in a solar system (made up of the sun, planets, moons, asteroids and comets) that sits about two-thirds of the way out from the galaxy's centre, towards the outer edge of the Orion Arm, made up of millions of stars and clouds of gas. Circling the galaxy at 828,000 kilometres per hour, our solar system makes one complete orbit of the Milky Way every 230 million years.

So how did we come to know all this just from looking up at the band of stars that arcs across our night sky?

It was William Herschel, working in his garden with home-built apparatus, who made the first systematic attempt at figuring out the shape of the Milky Way. He was a German-born English astronomer who lived and worked in Bath in the UK with his sister Caroline, who was a successful astronomer in her own right. William became world-famous overnight after he discovered the planet Uranus quite by accident when he was looking for double stars.

In 1785 Herschel presented a sketch to the Royal Society of London: his first attempt at the overall shape of the Milky Way. His methods were simple (too simple, it turned out) but his groundbreaking map gave us our first three-dimensional picture of our galactic home.

Knowing the distances to the stars from Earth is crucial, because if you know their distances and directions you can map out the structure of the Milky Way. Lacking this vital information, Herschel had to adopt some simple assumptions in order to make his drawing. First, he assumed all the stars were the same brightness as our sun, a simplification that enabled him to hazard a guess at the stellar distances.

Herschel picked the 600 or so brightest stars that he could see in all directions from Earth from both the Northern and Southern hemispheres. Using the inverse square law mentioned in Chapter 1 (which describes how bright an object would seem if you put it far away), he worked out the distance to each of the stars.

The result? The first sketch of our Milky Way. It showed a flattened disc of stars, slightly fatter at the middle, with the sun lying around a third of the way out from the centre. Not bad for a first attempt. Herschel didn't find the spiral arms, but mapping out the Milky Way's shape was still an extraordinary achievement for an eighteenth-century guy with a telescope in his garden.

Herschel's assumption that all stars are exactly as bright as the sun is actually not true. In fact, stars vary dramatically in brightness.

The faintest M-type dwarf stars shine orange with a feeble luminosity 1000th that of the sun. The brightest stars, O- and B-type supergiants, shine between 10,000 and a million times more brightly than the sun. This variation was unknown to Herschel and means we needed better ways to find the distances to the stars.

Fortunately, improved equipment and methods have since dramatically transformed our understanding of the Milky Way's size and structure. These tricks enable us to find the distances to stars and gas clouds throughout the Milky Way, to use them as markers of the spiral arms and other features.

Since its formation around 4.6 billion years ago, Earth has orbited around the sun, taking 365.25 days to make one full circle. A calendar year is 365 days, so the extra quarter of a day is added on every four years as 29 February—those years are known as leap years. The ceaseless motion of its orbit gently rocks Earth from side to side in space every six months, causing each and every star to nod very slightly from side to side on a six-monthly basis, an effect called *parallax*. This cosmic waggle dance is most noticeable when we look at the stars that are closest to us. To understand why, hold your hand about 5 centimetres in front of your nose with your palm facing your face. Point your index finger directly up to the ceiling and fold down your other fingers and thumb to form a fist. Looking directly at the tip of your index finger, close your left eye. Now, open your left eye and close your right. Switch back and forth between your eyes a few times. Notice how your finger jumps across your field of vision when compared to more distant (background) objects? Now try the same thing with your finger at arms length. What do you notice about how far your finger jumps across the field of distant vision?

By using the equations of trigonometry (the sines, cosines and tangents that probably bored you at school), we can use this nodding motion to measure the distance to each star. The distance to a star is equal to 1 divided by the angular shift in the star's position that occurs over a six-month period. The star's distance is measured in parsecs, a unit of measurement used by astronomers that is equal to 30 trillion kilometres, or 3.26 light years. The angle is measured in 'seconds of arc', which is a measurement of small angles equal to 1/3600th of a degree. The measuring units might sound complicated, but parallax is quite easy to measure in practice, at least for the nearest stars.

Although 1/3600th of a degree is a tiny angle and very difficult to measure using something large and clunky like a sextant, astronomers have developed modern techniques over the years to help. We now use arrays of radio telescopes spread widely across continents and linked by accurate atomic clocks (synched to an accuracy of a trillionth of a second) to measure the positions of stars and gas clouds to within three-billionths of a degree. We also use spacecraft, such as the European Space Agency's *Gaia* observatory, to precisely measure the positions of thousands of millions of stars. *Gaia* uses a fixed pair of telescopes, repeatedly switching its observing target to measure parallaxes with an accuracy of up to 500-millionths of a degree. Nowadays, we can measure the parallax of stars and gas clouds at distances of up to 36,000 light years using this technique. Earth's orbit around the sun enables us to measure the distances to stars as far away as the centre of our galaxy with an accuracy of 10 per cent. See? Trigonometry does have its uses!

Parallax is great for measuring the distances to stars in the nearby regions of our galaxy, but for more distant stars, the rocking motion over a six-month period is too small to be measured. A second way to measure astronomical distances is by using a type of star called a *Cepheid variable*.

The importance of periodic variable stars in measuring distances came to light in 1908 when US astronomer Henrietta Leavitt made an important breakthrough. She was working at Harvard College Observatory studying thousands of variable stars in the Magellanic clouds (two small galaxies that orbit the Milky Way). From her records, she realised that the fainter variable stars varied quickly, whereas the brighter variable stars were more likely to vary slowly. Furthermore, she determined a precise mathematical relationship between the period and luminosity of these variable stars, which came to be called Leavitt's law. This was a very important discovery, because by knowing the absolute brightness of a star (its luminosity) and measuring how bright it looks to us, we can figure out how far away it is, just as a light bulb appears bright if it's shining in your face, but from 2 kilometres away you'd barely notice it.

Other types of periodic variable stars were discovered soon after. Leavitt gave astronomy a new tool, the standard candle, which enables us to measure objects at distances that parallax can't reach. Because they are visible at great distances, we can use periodic variable stars to measure the distances to the far reaches of the galaxy to work out its total size. That's exactly what US astronomer Harlow Shapley did in 1918 when he measured the distances to several tight-knit groups of stars called *globular clusters*, which collectively form a spherical halo around our galaxy. By doing so, he estimated the diameter of the Milky Way to be around 300,000 light years—ten times bigger than everyone else believed at the time.

In the early twentieth century, the race to understand the scale of the universe was hotting up and Cepheid variables were right in the middle. It all centred around the nature of *spiral nebulae*—fuzzy blobs of light in the night sky that under the magnification of a telescope showed a spiral shape. New techniques that studied chemical fingerprints in the light from these nebulae suggested they were made up of millions of stars. Still, most astronomers assumed that the Milky Way was essentially 'all there is' and that spiral nebulae were just clouds of gas and stars living *inside* the Milky Way.

Shapley insisted that the spiral nebulae were inside our galaxy, but Heber Doust Curtis and many other astronomers believed that the spiral nebulae were galaxies in their own right, made up of thousands of millions of stars and separated from the Milky Way by vast spans of intergalactic space. That argument was preposterous, according to Shapley, because if the spiral nebulae were other galaxies, their apparent sizes would put their distances at hundreds of millions of light years away. If spiral nebulae were that distant, it would mean that the stellar novae seen in spiral nebulae were shining brighter than 10,000 million stars.

Sound outlandish? As it turns out, both of these things are true. Other galaxies lie at distances of millions of light years from our own. The bright 'novae' seen in spiral nebulae are actually supernovae that shine as brightly as the rest of the stars in the galaxy put together. Truth is sometimes stranger than fiction.

It was Edwin Hubble, another US astronomer, who finally resolved the debate over the location and size of spiral nebulae. In 1923 he was studying a large spiral nebula called the Andromeda spiral. Taking several pictures spread over time, he noticed that one of its stars seemed to vary in a regular way. He scrawled *VAR!* on the photographic plate in large letters, realising it could be a Cepheid variable star. He was excited because the presence of a Cepheid variable in the Andromeda spiral meant he could measure its distance. Plugging in the numbers, he calculated the distance to Andromeda to be 900,000 light years.[3] This was enormous in comparison with the size of the Milky Way, which at the time was understood to be 100,000 light years in diameter. It was the first clear evidence that Andromeda was another galaxy outside our Milky Way.

This was massive news. Before Hubble's measurement, it was thought that the entire universe was inside the Milky Way. Suddenly we realised that the spiral nebulae that filled the night sky were 'island universes', each separated by at least one million light years. Hubble, using Cepheid variable stars, had radically altered our view of the universe.

Once the 'spiral nebulae' were confirmed as distant versions of the Milky Way, there was natural speculation that the Milky Way also had a spiral structure. Astronomers began to hunt for the very brightest blue supergiant stars (which live primarily in the

3 We now know the Andromeda galaxy to lie almost three times this distance away, at 2.4 million light years, but that doesn't change the conclusion that the Andromeda spiral is another galaxy.

spiral arms of the galaxy) and the hot bubbles of gas that surround them, which are visible across large distances. These stars have a particular chemical fingerprint that allows their intrinsic brightness to be estimated. As a consequence, they can be used to determine distances, at least roughly.

Using this technique and a sample of just forty-nine stars, the first map of the nearby spiral arms was made in 1951 by William Wilson Morgan, whose announcement at a meeting of the American Astronomical Society was so exciting that it reportedly resulted in a standing ovation and stomping of feet!

At the same time as Morgan's discovery was announced, a team of radio astronomers were making strides towards understanding not just the nearby spiral arms but the structure of the galaxy as a whole. The technique they used was smart and powerful and is still used today to understand the spiral arms of the Milky Way and other nearby galaxies.

Our galaxy contains huge amounts of gas, the vast majority of which is hydrogen. Interstellar hydrogen is made up of very simple atoms that contain a proton in the middle and a single electron orbiting it. Protons and electrons have a property that is described by quantum mechanics as 'spin': they spin either up or down. You can imagine them as a basketball spinning around either clockwise or anticlockwise.

Most of the time, the proton and electron have the same spin, but once every eleven million years on average, completely at random, the spin of an electron will 'flip' 180 degrees to the opposite direction. This sounds a little wacky, but quantum mechanics tells us that fundamental particles can shift in this way. When the spins of the proton and electron flip from the opposite direction to the same direction, a tiny bit of energy is released as a radio wave at precisely 1420.405751 megahertz.

Although this only happens in any given atom once every eleven million years, there are countless trillions of hydrogen atoms in our galaxy, so it is actually a common occurrence and the sky is awash with radio waves.

These radio waves are an excellent tool for astronomers, since they give us an accurate record of where all the gas is in the Milky Way. Hydrogen is pretty much everywhere in our galaxy's disc, making it a great signpost of the structure of the Milky Way. Unlike light, radio waves can penetrate the clouds of dust and gas in our galaxy, making much more of its structure visible to us. We can also measure the speed of the hydrogen gas as it rotates around the centre of the galaxy, using a little trick called the Doppler effect.

The Doppler effect is the change in the wavelength of a wave (a sound, light, radio or other wave) caused by the relative motion of the source or the observer. With sound waves this translates to a change in pitch, which you notice when an ambulance screams past with its sirens on.

Doppler radar is used by meteorologists to measure the motions of clouds, telling us wind speeds and the direction of fronts and current rainfall. Waves are sent out by a transmitter on the ground and they bounce off the clouds, returning to the ground receiver station. As they reflect off the clouds, the waves change their wavelength depending on whether the clouds are moving towards or away from the radar station. Using the mathematics of the Doppler effect, meteorologists can build up a detailed picture of the clouds in three dimensions, measuring rainfall in real time.

As our galaxy rotates, astronomers use a very similar technique to measure the 'Doppler radar' map of the entire galaxy. Instead of transmitting radar from Earth (this would be too weak due to the dilution effect of radiation in space), we measure the radio waves that are produced at 1420 megahertz in clouds of hydrogen in space. The wavelength used for Doppler radar is much shorter than the radio waves emitted by hydrogen atoms in space. That's just as well, because in space we need the radio waves to sail straight through the clouds to reach our telescopes on earth. Long-wavelength waves pass easily through objects such as clouds, whereas shorter waves are reflected. If we were looking at shorter-wave radio signals, the waves would reflect off each interstellar cloud and never reach us at all.

Astronomers have employed this technique using radio emission from hydrogen and carbon monoxide (which is also ubiquitous in the spiral arms) to produce Doppler-radar-style maps of the Milky Way. What we see is a series of spiral arms wrapping around the galaxy in a logarithmic spiral. This shape is also seen in nature—for example, in the shells of snails, in ferns, and in cloud bands in hurricanes.

The sun lies in the Orion Arm, between the Sagittarius and Perseus arms. Unlike the major spiral arms of the Milky Way, which are long and slender and wrap around the galaxy, ours is a minor protuberance whose origin is not really well understood. It is most likely due to a disturbance in the galaxy's standard four-armed structure, perhaps related to a past collision with a smaller galaxy or a high-velocity cloud (a vast cloud of hydrogen seen above or below the galactic disc, travelling at several hundreds of kilometres per second).

The Doppler technique is not perfect. From our vantage point in the Milky Way's disc we are seeing several arms, one behind the other. That makes it quite hard to distinguish one from the other. Studying the spiral arms in the shadow of the giant 'bulge' of stars and gas situated around the galactic centre is also difficult. But by combining the Doppler technique with measurements of the motions or parallaxes of individual stars and star-forming regions, it has been possible to map out at least four distinct spiral arms that wrap around the galactic centre.

The spiral arms are not physical structures that rotate around the galaxy. If they were, they would quickly get very tightly wound because the gas towards the edge of the galaxy is orbiting more slowly than the gas towards the middle. Within a couple of orbits the arms would wrap completely around the Milky Way, merge and destroy themselves. Most astronomers believe that the spiral arms are caused by density waves—think of traffic jams around an intersection—leading to compressions in the gas of the Milky Way's disc that trigger star formation as the gas is compressed.

As the density waves pass, gas in the spiral arms is compressed and pockets of compressed gas become more concentrated and collapse under gravity. This forms large numbers of young, hot stars and explains why the spiral arms are blue—they contain a large concentration of young hot massive stars that are very blue in colour because of Planck's law, which states that the colour of light emitted by an object depends on its temperature. Red-hot coals are less than 100 degrees Celsius, whereas a blue gas flame is around 2000 degrees Celsius. In contrast, the central bulge of the Milky Way is red, because it is undisturbed by density waves and contains only older stars, which turn redder as they cool over billions of years.

The centre of our galaxy is also a place of great extremes. The so-called Central Molecular Zone (CMZ) is a pocket of the inner galaxy stretching 700 light years. It contains almost 80 per cent of all the dense gas in the galaxy—a reservoir of gas tens of millions of times the mass of our sun, where complex chemicals build up on grains of interstellar dust under the enzymic action of ultraviolet radiation. Surveys with radio telescopes reveal colossal stores of ammonia, carbon monoxide, silicon monoxide, water and even larger chemicals that have come about by the build-up of heavier elements in the cores of stars and the subsequent release of these chemicals into space by supernova explosions.

The CMZ is denser, warmer and more turbulent than similar molecular regions in the disc of the Milky Way but, strangely, is producing ten times fewer new stars than would be expected from the sheer volume of gas. I'm studying this fascinating region of our galaxy with my colleague Dr Maria Cunningham and two PhD students, Shaila Akhter and John Lopez, at the University of New South Wales. We are using a 22-metre-diameter radio telescope nestling in a valley on the edge of the Warrumbungle National Park to study very dense and cold clumps of gas believed to be the seeds of new stars. These stellar nurseries are moving through space at unusually high speeds, suggesting that the gas is being stirred up by the *galactic bar*, a cigar-shaped arm of material lying straight across the centre of the Milky Way. We believe that the turbulent conditions created by the galactic bar could be inhibiting the formation of new stars, which would explain the dearth of new stars currently being assembled in the CMZ.

Deep in the heart of the galaxy, at the very core of the Milky Way, hides a supermassive black hole called Sagittarius A★ (pronounced A-star) that is thirty times the diameter of the sun and harbours four million times its mass. A *black hole* is an accumulation of matter so massive and compact that the very fabric of space and time around it forms a deep 'well' from which nothing (not even light) can escape. The gravitational field of the object is so strong that it rips the 'surface' of space, thus inhibiting the passage of light.

We first came to suspect something unusual was living at the centre of the Milky Way when an extremely bright beacon of radio waves was noticed in the Sagittarius constellation in 1974. The source of the emission was extremely compact and seemed to coincide precisely with the direction of the centre of our galaxy. After monitoring it over the course of a few years, radio astronomers used the parallax technique to prove that Sagittarius A★ lies smack-bang at the centre of the Milky Way.

The realisation that Sagittarius A★ was a black hole came sometime later. Its mass was first estimated by tracking the orbit of a star called S2 over a period of a decade. This was possible because knowing the period and diameter of S2's orbit enabled us to calculate the mass of the central object. Having a swarm of orbiting objects constrained the mass even more accurately. Astronomers have now been following the orbits of more than a dozen stars around Sagittarius A★ for more than fifteen years, narrowing down the mass of the black hole to 8.5 trillion trillion trillion kilograms, or 4.3 million times the mass of our sun.

So what does it look like? Most people imagine a black hole to be completely dark, but that is not the case. Although no light escapes the black hole's *event horizon* (the boundary of no return for rays of light), there is no shortage of radiation emanating from the active region just *outside* the black hole. The bright radio emission from Sagittarius A★ is thought to come from a region of gas called an *accretion disc*—a pancake-flat cosmic waiting room for material that is swirling around the black hole, waiting to fall in. As material is gradually swallowed by Sagittarius A★, more stars and gas are pulled onto the disc and it replenishes itself. The diameter of the Sagittarius A★ black hole is around 88 million kilometres, the same as the orbit of Mercury (the innermost planet) around the sun. Both the black hole and its accretion disc are completely invisible in optical light, thanks to the huge amounts of dense light-absorbing clouds that lie between us and Sagittarius A★.

We don't know how Sagittarius A★ originated, but several theories have been developed by theoretical astrophysicists. One hypothesis is that it formed shortly after the birth of our galaxy some twelve billion years ago, when a large number of massive stars went supernova and created a swarm of black holes that gradually accumulated and amassed into one supermassive black hole at the centre of the galaxy. Another theory is that the supermassive black hole was created by the direct collapse of enormous volumes of gas in the early stages of the universe. Although both of these scenarios are theoretically possible, we currently have no idea which (if either) is true.

Recent efforts to understand the behemoth at the galactic centre have focused on studying a fleet of stars that are orbiting the black hole. This group of twenty-eight stars takes between sixteen and 1700 years to complete one orbit, moving at supersonic speeds of up to 2500 kilometres per second. In precisely the same way that the planets orbit the sun, the vast majority of the stars circle Sagittarius A★ in stable elliptical orbits, at no risk of falling into the black hole's jaws. The elliptical motion is a finely tuned balance between the motion of the star through space and the gravitational dragging of the star towards the black hole. With forces balanced so precisely, the star is in no danger of being swallowed up.

As you get closer and closer to the event horizon of the black hole, the gravitational force becomes far stronger. In fact, by halving your distance to the black hole's edge, the forces on a body quadruple. In this zone, stars are literally ripped apart as their gas is pulled and stretched by the strong gravitational fields close to the black hole's event horizon. When this happens, some of the gas streams onto the accretion disc—the white-hot carousel of material awaiting its final journey into oblivion. Although we have never witnessed a star being ripped apart by the supermassive black hole at the centre of the Milky Way, we do observe these events from time to time in other galaxies, with gigantic bursts of X-ray, ultraviolet, optical and radio waves produced by the searing-hot cauldron of gas signalling a star's final demise. And with so many different types of galaxy in our universe, many of them hosting supermassive black holes at their centre, there is no shortage of these fascinating objects to study.

3

A GALAXY MENAGERIE

As they gazed into the brilliant night sky, our ancestors had no idea about the riches that lay unseen. As they shared stories around campfires or devised tales of sentient constellations to explain the world, they remained oblivious to the estimated 100 billion galaxies in the universe, which come in a dazzling array of shapes and sizes. Now, in the era of the astronomical telescope, we have photographed literally hundreds of millions of them in vast surveys using Earth-orbiting optical and infra-red telescopes and giant radio telescopes. Like butterfly collectors, we sort and classify them into species and observe their environments to learn more about their lives, their behaviours and their interactions.

We search for rare gems—the platypus and the toucan, the weird and wonderful types. We witness snapshots of galactic collisions between grand spirals, and the creation of brilliant fields of new stars. We see the battered shells of mangled hybrid galaxies and piece together the lives of elliptical galaxies—the bloated elders of the skies. This amazing galaxy menagerie is giving us new insights into the creativity of the cosmos. Scientific dissection of these samples is beginning to uncover the circuitous path trodden by these giant cradles of stars as they form, interact and transform throughout their billions of years of life.

You and I live in a pretty average example of the most common type of galaxy: the spiral. A single image from the Hubble Space Telescope in any direction reveals hundreds of spiral galaxies, some seen edge-on and appearing thin and serpentine, others proudly displaying their slender arms like an elegant octopus. Spiral galaxies range from around 10,000 to 500,000 light years in size and tumble in all directions in space. Their constituent stars and gas rotate in a near-circular motion that, like pizza dough tossed into the air, causes them to flatten into a thin disc.

Galaxies are classified according to their apparent shape and features. We still use a classification scheme that was devised by Edwin Hubble in the 1930s, albeit with an increasing number of embellishments as new types or subclasses of galaxies are discovered. In this scheme, spiral galaxies are classified as Sa, Sb or Sc, with Sa having the most tightly wound spiral arms and Sc being the most relaxed.

We can tell a lot about a galaxy by looking at the colour of its stars. As the density wave in the spiral arm compresses the gas, gravity is helped along in its quest to clump the gas together in smaller pockets. These clumps of gas become hot and dense and eventually become stars. The passage of a compression wave tends to create a lot of very massive stars that are extremely luminous and dominate the light output of the spiral arms. Hot objects emit more blue light than red (that's down to Planck's law of quantum mechanics), so spiral arms with active star formation tend to be blue.

Regions of the galaxy that are devoid of active star formation tend to have fewer massive stars and appear redder. The red colour comes from cooler stars, which emit all colours but are brightest towards the red end of the spectrum. Of course, no stars are really 'cool'. Red stars have surface temperatures of more than 3000 degrees Celsius, but that is nothing compared with a hot massive blue star, whose surface temperature comes in at a snuggly 30,000 degrees Celsius.

You can easily see the colours of the stars with your unaided eye. Consider the constellation of Orion, for example: it is bright and easily visible from pretty much everywhere on Earth. Take a look next time you get a clear night and you will see that almost every bright star in Orion is blue—that is, except for one star in the corner of Orion's rectangular body that is bright orange. You'll find it at the bottom-left corner of Orion if you live in the Southern Hemisphere, or the top-right corner if you're in the Northern Hemisphere. This is Betelgeuse (often pronounced 'beetle-juice'), a fantastically bloated red supergiant star nearing the end of its life. Betelgeuse is 1000 times larger than the sun—in fact, if the sun were replaced by Betelgeuse its surface would reach as far as the orbit of Jupiter, and Earth would be totally consumed!

If you zoom out far enough, spiral galaxies look like (to paraphrase the late English amateur astronomer Patrick Moore) 'two fried eggs clasped back-to-back'. Their flattened discs are surrounded by a spherical bulge of cooler stars concentrated towards the centre. Galactic bulges come in two flavours. The classical type contains very old stars (typically more than eight to ten billion years old) that orbit the galactic centre in random directions. They contain little or no gas from which to form new stars, and therefore appear quite red. The other type of galactic bulge, discovered more recently using powerful instruments like the Hubble Space Telescope, is more like a mini spiral galaxy within a galaxy. In this type of bulge, the stars generally rotate in the same direction as the galaxy's disc. These nested bulges contain gas clouds and maintain an active program of star formation, making them appear bluer than their spherical counterparts.

Many spiral galaxies, including the Milky Way, have a straight arm called a 'bar' of stars and gas lying across their middles. This always prompts the joke:

Q. Where does an astronaut go to drink?

A. The space bar.

In a barred spiral galaxy, the spiral arms begin their curved paths at each end of the bar. Although the angular shape of a bar might seem out of keeping with the slender curves of a spiral arm, computer simulations of galaxies show that bars will readily form if left to their own devices. They seem to form slowly over time—in a recent study of more than 2000 face-on spiral galaxies, around 70 per cent were estimated to have a bar, but when the universe was half its current age, that figure was closer to 20 per cent. If galaxies are disturbed by interactions with other galaxies, on the other hand, the bars are often disrupted or destroyed.

The presence of a bar seems to affect a galaxy in two ways. First, it can trigger star formation towards the galactic centre. As the galactic bar rotates it acts like a spoon stirring coffee, driving the galaxy's gas and dust towards its middle and creating stars in its wake. This has the effect of building up the galaxy's bulge and removing star-forming fuel from the disc, which in turn reddens the spiral arms. Recent research shows that redder galaxies are far more likely to have bars—which corroborates this hypothesis.

It might sound contradictory, but the second effect that this stirring can have is to move some of the gas around in strange orbits. This gas can then pool towards the ends of the galactic bar and the constant stirring motion inhibits star formation in these regions. A study undertaken in 2017 by my PhD student Shaila Akhter reveals that two large gas clouds near the two ends of the Milky Way's bar are far more turbulent than similar regions in the inner galaxy. The high random velocities of the gas seem to have inhibited star formation in these regions, possibly by preventing the dense clumps of gas within the clouds from collapsing to form stars. We don't know if this snapshot view of our inner galaxy is representative of all galactic bars, but one thing is for sure, these galaxy-wide processes are complex. The various effects of the bar on star formation are still under investigation.

At the centre of most galaxies, at the heart of the galactic bulge, lies an invisible object that is around four million times heavier than our sun. It picks up hundreds of errant stars, spins them around like a wizard and wears them as a cloak. If one of the stars dares to get too close, it is ripped apart and devoured in a blaze of fiery sparks. What's more, this nuclear ninja is incredibly dense—its precise diameter is not known for sure, but at radio frequencies its size has been determined to be less than the distance between Earth and the sun. Furthermore, the object has been monitored for more than a decade, and unlike every other object in the galaxy it has almost zero motion through space. It just sits there. What cosmic creation could behave in such a way?

We can determine its nature by a simple process of elimination. Could it be gas?

Well, the answer here is a definite no. Interstellar gas has a density of one atom per cubic centimetre, so if you filled Earth's orbit around the sun with this gas you'd have around 5 trillion kilograms of gas, which is a trillion trillion times short of the mass required to cause the stars' motions.

How about stars?

Still not even close. If we crammed more than a million stars into a space the size of our solar system, they would be quickly overtaken by mutual gravitational attraction and collapse into a heap. The densest collection of stars we've ever seen, a globular cluster, falls short of the density required by 1000 trillion times.

By a process of elimination, we reach our most probable cause for the motions of stars around the centre of the galaxy: the only physical object in the universe known to have this much mass crammed into a tiny region is a black hole. Supermassive black holes with millions or even several billions of times the mass of the sun are now thought to reside in the centre of almost every galaxy in the universe.

Beyond the black hole, disc and bulge, spiral galaxies sport a large spherical 'halo' of stars, star clusters and hot gas. Imagine a delicious breakfast roll. If the egg white is the galaxy disc and the yolk is the bulge, then the halo is the bun. *Galactic haloes* contain millions of lone stars that are whizzing through space at hundreds of kilometres per second. For a long time astronomers believed that these lone rangers were distributed randomly, but a closer look at the Milky Way with large optical and infra-red telescopes showed that many of them actually huddle in streams or ribbons displaced from their home galaxy. These ordered subsets of the halo, called galactic *streams*, are oriented at a range of angles to the Milky Way's disc and a few of them make a complete orbit around our galaxy. Eighteen galactic streams are currently known around the Milky Way, and their orbits have been calculated with some precision.

The Sagittarius stream, which contains many millions of stars, wraps around our galaxy at nearly 90 degrees to the disc. It links us to the Sagittarius dwarf elliptical galaxy, a collection of middle-aged stars that is thought to be in the process of merging with the Milky Way. It orbits us once every billion years, and as it periodically intersects the Milky Way's disc the smaller galaxy punches a hole, leaving a river of stars trailing in its wake.

Not all galactic streams are made of stars. The Magellanic stream that links the Milky Way to the large and small Magellanic clouds (two dwarf galaxies that are gravitationally bound to the Milky Way) contains more than 200 million times as much hydrogen gas as is contained in the sun. It contains no stars at all, except in an isolated region where the leading edge of the Magellanic stream is interacting with gas in the Milky Way's disc, triggering a burst of star formation.

Galactic streams are not unique to the Milky Way. They are also seen in nearby galaxies such as the Andromeda galaxy, which lies 2 million light years from Earth. Looking at the orbits of galactic streams makes it clear that they are caused by small satellite galaxies or globular clusters being shredded by gravitational forces as they venture too close.

Galactic haloes aren't just made up of lone stars and gravitational tidal streams. They also contain swarms of hundreds of thousands of stars densely packed into balls; we call these *globular clusters*. With the naked eye a globular cluster looks like a faint, fuzzy ball of light, but through a pair of binoculars or a small telescope you can clearly see that it is made up of countless pinpoints of light.

Omega Centauri is a globular cluster that is an excellent target for backyard astronomers. You can spot it as a fuzzy star close to the Southern Cross. First find the six stars that appear on the Australian and New Zealand flags—the four stars of the Southern Cross and the two 'pointer' stars to its left-hand side, called Alpha (left) and Beta (right) Centauri. Now find the imaginary middle point of this group of stars, between the cross and the pointers. Move upwards, about twice the height of the cross, and you will find a rather faint but fuzzy object. This is the globular cluster Omega Centauri, which lies 16,000 light years from Earth.

Zoom in with binoculars or a small telescope and you'll see what looks like a distant swarm of bees. The ten million stars that make up the cluster fill a region approximately 150 light years across, making it the largest globular cluster in the Milky Way. I first saw Omega Centauri from the Sutherland Astronomical Society's telescope at the Green Point Observatory near Cronulla in south-east Sydney. Even with the light pollution in this sprawling city of four million people, the sight took my breath away.

When we look at a globular cluster through a professional astronomical telescope, swapping the camera for an instrument called a spectrograph, we can measure the chemical make-up of the stars from the light they emit and absorb from atoms in their atmospheres. Doing this, we see that stars in globular clusters have very low levels of elements heavier than hydrogen and helium. Why is that important?

At the beginning of the universe, everything was made from hydrogen and helium. The so-called 'heavier' elements (ones further along the periodic table, such as carbon, nitrogen and oxygen) were generated only by the last one to three generations of stars. The lack of heavier elements in globular clusters therefore points to them being very old. In fact, we think that most globular clusters are around twelve billion years old, which means that they formed very early in the lifetime of the universe.

Globular clusters are also devoid of gas and as a result are unable to form new stars. We have no evidence of star formation occurring in any globular cluster. Presumably all the gas was used up in star formation long ago. This makes them unique environments for studying how older stars behave.

Recent (and still controversial) evidence has come to light that some globular clusters have something very big lurking at their centre. Motions of stars in several globular clusters, including M31 G1, 47 Tucanae, NGC 6624, NGC 6388, NGC 6752, M15 and Omega Centauri, have been studied to determine how big and heavy these objects really are. What researchers have found is that the weird motions of the stars could be explained by the presence of a central black hole with thousands of times the mass of the sun—a so-called *intermediate-mass black hole*.

This rare and previously hypothetical flavour of black hole has been on our radar for a long time and is seen as a possible 'missing link' between black holes created by a single star, which are scattered throughout galaxies, and supermassive black holes, which have millions of times the mass of the sun and live in galaxies' centres. The amalgamation of intermediate-mass black holes over billions of years is a possible explanation for the origin of supermassive black holes. If their existence is verified, it would be a very significant step towards understanding the evolution of black holes and, indeed, the evolution of galaxies.

The trouble is that evidence of intermediate-mass black holes is spotty at best. It's true that a handful of cases have been found where the motions of stars or gas could potentially be explained by the presence of a compact mass with thousands of times the mass of the sun. However, credible alternative explanations for the dynamical motions of stars in these clusters have subsequently been put forward for many cases, and the argument for intermediate-mass black holes, although exciting, has failed to convince the majority of astronomers. We still await incontrovertible evidence of intermediate mass black holes.

Despite our lack of definitive proof that intermediate black holes exist, the busy centres of globular clusters would be ideal places to host them. Stars in globular clusters are packed on average 1000 times more densely than stars in our solar neighbourhood—each star in a globular cluster is approximately 1 light year from its nearest neighbour. The densely packed cores of globular clusters are even more crowded, holding on average one star within a volume equivalent to that of our solar system. In these cosy environments, collisions between black holes would not be out of the question. The accumulation of small black holes to form intermediate ones would be very likely in such a congested environment.

Similar processes certainly happen to stars in the crowded cores of globular clusters. These regions are home to a handful of weird and wonderful stars called *blue stragglers*, so-called because they are far hotter and have two to three times more mass than their neighbouring stars. The most widely accepted theory is that blue stragglers are the result of a star-on-star collision and merger—a direct hit in an unusually crowded environment. Imagine playing a game of snooker where two red balls collide and they turn into a blue. That seems to be what has happened to these poor old blue stragglers.

Given that globular clusters are almost twice as old as our solar system and the stars in them live very long and stable lives, some astronomers have suggested that they are prime candidates for searching for signs of technological extraterrestrial civilisations. The argument goes like this: within their twelve-billion-year lifetimes, stars in globular clusters have plenty of time to develop life, which would have ample time to generate intelligence, then complex technology and possibly interstellar travel. Interstellar travel is much easier in a densely packed cluster of stars, because the alien beings would be able to travel between neighbouring stars within a few Earth years.

By targeting globular clusters in our search for signals from alien technologies, we might just stand a stronger change of finding something. If we assume that the aliens' interstellar communication relies on electromagnetic waves (for example, light or radio waves), and that this communication is capable of reaching other stars, it is possible that our most powerful telescopes may be able to detect these signals. It's not a bad idea in theory—after all, if we put our most powerful radio telescope on a planet around Proxima Centauri, our nearest star after the sun, we would be able to detect human presence on this planet from all the airport and weather radars that we beam into the sky on a daily basis.

There are a few (mainly privately funded) groups of astronomers around the world who are actively searching for signals from technological civilisations in our galaxy. Attempts have been made by these ambitious dreamers to listen for radio signals from alien civilisations coming from globular clusters. In the late 1970s, a search was conducted of twenty-five globular clusters, encompassing a total of ten million stars, at two specific frequencies emitted by hydroxyl and water molecules. Since these are frequencies that arise by a common natural process in space, the astronomers involved figured that aliens might expect that we would be looking for signals at those frequencies and therefore crash the party and transmit brightly at those frequencies.

How would we distinguish natural signals from alien ones? Perhaps they would flash on and off in some sort of code, or perhaps the aliens would create unnaturally bright signals in the hope that someone would see them. They could just be going about their business, communicating with their own fleet of spacecraft, and we are the eavesdroppers. That is the hardest thing about the search for extraterrestrial intelligence—second-guessing how they might communicate and whether they want to be seen at all. We just don't know. Nothing has been found in more than fifty years of searching.

We have also tried to talk to aliens, making our presence known in a somewhat controversial experiment. In 1974, astronomers beamed the most powerful broadcast ever from the Arecibo radio telescope towards the globular cluster M13, which lies 22,000 light years from Earth. The message was three minutes long and comprised an encoded binary radio transmission from the megawatt antenna at Arecibo. It showed a blocky pixelated image of the Arecibo telescope, a visual representation of our solar system with Earth poking up to indicate where we are, and then a stick figure of a human with no head (I never did understand that part), DNA strands, and a complex series of numbers representing some of the chemicals in DNA.

Some people at the time were livid. How dare these scientists endanger our entire species by giving away our presence on Earth! To be honest, I don't think we should be too worried, since we can't expect a reply (or a visit!) any time soon. The signal, which travels at the speed of light, won't arrive at M13 until the year 24,000 CE and is only three minutes long, so if they blink they will miss it. Even if they do reply, or strap on their space boots and decide to pop by for a cuppa, their message (or their spaceship) won't reach us for at least another 44,000 years.

Most current searches for ET are focusing on radio waves coming from planets around the nearest stars to Earth. These are a good solid bet for finding accidental relatively weak radio signals from things like communication and TV or radio broadcasting towers. They also hold the best hopes for actually *communicating with* a civilisation, because the light-travel time between Earth and the alien planet would be shorter than a human lifetime. Globular clusters are much further from Earth, making them poor candidates for a search for weak signals. On the other hand, their environment and the sheer number of stars we can look at with a single observation make them excellent places to maximise our chances of finding something if the signals are strong enough to travel vast tracts of space. This assumes that we don't care about taking 50,000 years to exchange greetings! It's all a guessing game at this stage.

Although the odds of finding something are incredibly low, it is beyond doubt that if we were able to tune in to a broadcast from an alien television station, it would be one of our greatest moments as a species. What would we do next? I think the best thing we could do would be to sit, listen and learn. If they're anything like us, they might just be dangerous.

One place that may not be ideal for hosting alien life is an elliptical galaxy. With no spiral arms, ellipticals are classified in Hubble's scheme as E0 (completely spherical) to E5 (long and thin). They're not exactly pretty—their main characteristic is, well, plainness. Elliptical galaxies have no spiral arms and very little interstellar gas floating around, and therefore most of them lack any significant ongoing star formation. They are basically one big galaxy bulge made up of old red stars with a supermassive black hole at the centre. Even the orbits of the stars are weird—they don't rotate in a lovely disc like the Milky Way, but in randomly oriented orbits.

Ellipticals have a large number of globular clusters that, like the globular clusters in a spiral galaxy, are arranged in a spherical halo. In fact, elliptical galaxies have two to three times the number of globular clusters as spiral galaxies of the same brightness. And there is something odd about the globular clusters in elliptical galaxies: they come in two distinct populations. One comprises older stars made of hydrogen and helium, and the other is made up of newer stars with higher abundances of heavy chemical elements. That's a pretty important hint about the origin of elliptical galaxies. At some stage, these new and chemically rich globular clusters have presumably formed within the elliptical galaxies, or have been acquired by mergers with other galaxies.

Another hint is supplied by the study of elliptical galaxies through the history of the universe. Looking back in time, using the fact that light travels at a finite speed and therefore the light from distant galaxies is only just reaching Earth, we can see that around ten billion years ago the numbers of elliptical galaxies rose rapidly. This is another clue that they may be pretty featureless but have led very interesting lives. The body of evidence suggests that the formation of elliptical galaxies is closely connected to collisions and mergers between smaller galaxies. These events were far more common in the past when the universe was a very crowded place, and that is what precipitated the explosion in the number of elliptical galaxies around ten billion years ago. As we'll see later, major transformations are not only possible but virtually guaranteed when galaxies collide.

Further clues to the role of collisions or mergers in the evolution of galaxies come from the many varieties of hybrid galaxy seen in the night sky. The main two are called lenticular (lens-shaped) and irregular.

A lenticular galaxy is something of a mix between a spiral and an elliptical galaxy. Denoted S0 by Hubble, lenticular galaxies have a large bulge of older, redder stars as well as a thin disc. The disc is almost featureless, having no spiral arms and therefore no large-scale star formation. Most lenticular galaxies contain almost no stars less than one billion years old. The reason for this lack of star-children seems to be that lenticular galaxies have somehow lost their interstellar gas. They do contain a lot of cold, dense molecular gas, however. This is the material that is spewed out into deep space by *supernovae*—explosions of massive stars at the end of their lives. Along with hydrogen, molecular gas can go into the mix to form new stars, but it cannot do the job of star formation without large quantities of hydrogen, since this is the fuel that enables a star to ignite.

Irregular galaxies, as their name suggests, have a range of weird and unexpected shapes. They can be featureless blobs or complex, mangled train wrecks. It's a broad-brush definition and can include weirdly shaped individual galaxies, including dwarf galaxies like the small Magellanic cloud that orbits the Milky Way, or remnants of cosmic crashes whose wreckage resembles none of the other galaxy types.

One of the most spectacular examples of an irregular galaxy is the Cartwheel, in the constellation of Sculptor. It looks like something out of a fireworks display. An image from the Hubble Space Telescope shows a small, reddish spiral galaxy surrounded by a large, glowing blue ring made up of bright young stars, with spokes linking the two. The Cartwheel is thought to be the result of a smaller galaxy passing directly through a larger one around 200 million years ago. The resulting shock wave blasted a ripple through the resulting mash-up and compressed the gas, triggering a wave of massive star formation propagating through the large blue ring that surrounds the surviving spiral.

Sorting galaxies into similar types has led to an improved understanding of our universe. It can be difficult to classify many galaxies accurately by eye. From our limited viewpoint on Earth, a long, thin elliptical galaxy looks virtually identical to an edge-on spiral galaxy. Only by travelling there (which would take millions of years in the best spacecraft imaginable) could we fly around and see the galaxy in three dimensions.

In recent years, we've photographed millions of galaxies using cameras on ground-based and Earth-orbiting telescopes in an attempt to better understand their formation and evolution. Due to the enormous increase in the capabilities of telescopes, which are now equipped with powerful digital cameras, there is simply no time for astronomers to sit and classify galaxies one by one. You might think that there is no need for manual classification—surely computers are now able to do a better job than people, especially with the advent of modern machine learning and shape recognition technology? Surprisingly, despite work to improve the accuracy of computer classification of galaxies, humans are still ahead of the curve when it comes to accuracy and the ability to identify, and in particular to identify and question when they see something unusual.

That's why galaxies are now increasingly classified by citizen scientists as part of projects like Galaxy Zoo. This free crowdsourcing project, which has been online since 2007, gives members of the public access to astronomical data for the purposes of classifying galaxies and discovering new and interesting objects. On the Galaxy Zoo site, you can access more than a million galaxies from the Sloan Digital Sky Survey (which uses a 2.5-metre-diameter optical telescope located in New Mexico) as well as images from the Hubble Space Telescope, which operates from Earth orbit, and the 3.8-metre-diameter United Kingdom Infra-Red Telescope, which sits above the clouds atop Mauna Kea in Hawaii. The project shows participants a series of real images of the sky, and the volunteers are asked a series of questions about the shape and characteristics of the galaxies that are visible in the pictures. The questions are designed to ascertain the type of galaxy (for example, spiral, barred spiral, elliptical), the shape and size of its bulge, and whether it is interacting or merging with another galaxy.

Galaxy Zoo was so popular in its first year that the team of professional astronomers who run the project received over fifty million galaxy classifications! What's more, the citizen scientists have made several significant discoveries, earning them places as co-authors on more than fifty peer-reviewed research papers. They have discovered a whole raft of blue elliptical galaxies, which look like ellipticals but contain young stars, and have also unveiled red spiral galaxies, containing mostly old stars.

One of the most exciting discoveries made by a Galaxy Zoo participant so far was in 2007, by a Dutch schoolteacher and citizen scientist called Hanny van Arkel. She was carrying out a routine classification task on an image from the Sloan Digital Sky Survey. The picture was clear, with a bright galaxy at the centre of the image, which she classified as an anticlockwise spiral. Then she noticed a strange object—a blue smudge of light hanging just below the galaxy. She used the feedback form to alert the astronomers at Galaxy Zoo to this space oddity, asking, 'What's the blue stuff below? Anyone?' Their response was: 'Interesting'.

When presented with an unidentified object in space, astronomers always rise to the challenge. The Galaxy Zoo astronomers quickly scrambled their available resources and were able to carry out further observations with the Chandra X-ray telescope and the Hubble

Space Telescope to find out more. By measuring the characteristics of the light from the strange object (by now called Hanny's Voorwerp, which is Dutch for 'Hanny's object'), they found that the light emitted by the cloud was a very close match to the light that comes from the core of an active galaxy when it is ejecting radiation and gas into intergalactic space in response to the consumption of an all-you-can-eat banquet by a massive black hole.

The most likely scenario for the faded glow that van Arkel had discovered, the researchers found, was that it originated from a gigantic stream of gas that had been extracted from the nearby galaxy by the close approach of a smaller, satellite galaxy. This gas tail was now being illuminated by the light that came from a recent outburst from the region surrounding a supermassive black hole in the nucleus of the galaxy, like a ghostly reflection of past glories. The enormous event that had disrupted the black hole also unleashed a galaxy-scale super-wind of gas and radiation that blasted into the Voorwerp and triggered a huge cluster of new stars to burst into life. Not a bad discovery for an amateur!

Hanny's Voorwerp shows us just how powerful supermassive black holes can be. But in most galaxies they lie dormant for millions of years before unleashing their power. When they do, this can switch on a highly disruptive phase of a galaxy's life, transforming it into an 'active galaxy'. We study active galaxies by pointing our radio telescopes at the night sky to reveal thousands of extremely bright dots of radio wave emission. These are not stars. Each one is the active nucleus of a distant galaxy—dramatic, powerful and visible across billions of light years of space.

Radio astronomers first discovered active galaxies in the 1940s using radar equipment left over from aircraft communications and surveillance operations in World War II. If you're ever in Sydney, walk north along the coast from Bondi Beach to the Rodney Reserve in Dover Heights. There you will see one of the first sites of radio astronomy in the world. I've visited this site on several occasions, and despite the fact that the equipment is long gone, it is always inspiring for me to be in a place where the science to which I have dedicated my career was born. The radio equipment there was originally designed to detect enemy warships and aeroplanes, but after hostilities ended it was used to study the variable radio emissions from sunspots and solar flares. Its crowning triumph was to discover some of the brightest active galaxies in our skies. These fire-breathing dragons had strange properties and were very poorly understood, and rapidly became a hot topic of research. You could say that the discovery of active galactic nuclei accelerated the development of radio astronomy and was one of the driving forces behind the decision to construct giant radio telescopes such as the Parkes radio telescope in Australia.

An active galaxy is dominated by *jets*—narrow streams of tiny subatomic particles spraying outwards from the centre at a rate approaching the speed of light. The particles are accelerated by the intense magnetic fields at the north and south poles of the accretion discs around the supermassive black hole at the core of the galaxy. The jets often end in 'lobes' of radio-emitting material, each shaped like an oyster mushroom, that stretch far outside the visible galaxy of stars and gas.

Centaurus A is a spectacular example of an active galaxy. From our vantage point on Earth we get a fabulous view because it is (relatively speaking) very close to the Milky Way. In an optical telescope it looks like a fairly normal galaxy with a bright sphere of older stars, a disc of blue stars and a dark stripe around its middle. We understand the stripe to be a band of molecular clouds left over from generations of stars and their supernovae. With a radio telescope Centaurus A is enormous—the jets are a million light years long and the lobes stretch more than 30 square degrees across our sky. If your eyes were capable of seeing radio waves, you would see this thing taking up 150 times the area of the full moon! That's quite a feature we're missing.

The jets of active galaxies don't just look spectacular—they also have a profound effect on their host galaxy. When jets form, they seem to put a rapid stop to star formation by blowing large amounts of cold molecular gas (the fuel for new star formation) out into intergalactic space. This was recently observed directly for the first time in the nearby galaxy NGC1266, where astronomers witnessed 20 million solar masses of cold gas being blown out of the galaxy at speeds of 500 kilometres per second, a rate that will see the gas all gone in eighty-five million years. Unless new molecular gas comes along from another source (for example, a merger with another galaxy), the gas purge will doom NGC1266 to a lonely retirement with no stellar nurseries and no new generations of stars.

The great diversity of galaxies raises the questions: how do such a wide range of galaxies form, and how do they evolve?

As we witness more and more of these dynamic phases of their lives, it has become apparent that a galaxy is not one type or another, not forever a spiral or an elliptical or an active galaxy. These are simply phases in its complex journey through its billions of years of life. Like living creatures on Earth, galaxies interact and merge with other galaxies. They create new stars, change throughout their lifetime and evolve to suit their environments.

With this breakthrough realisation, we can work towards understanding the evolution of galaxies over the past thirteen billion years, as well as predicting how galaxies (and the Milky Way in particular) will evolve in the future.

4

THE FORCE IS WITH YOU

The late Gene Cernan was the last human being to leave his footprints on the lunar surface, back in 1972. Flying from Perth in Western Australia in a chartered aircraft in May 2016 were me, Cernan and the cast and crew of a live stage show called *The Last Man on the Moon*. We were heading two hours north towards Carnarvon, a small coastal town, for our second show of a national tour I was hosting for a company called Live on Stage Australia.

Also along for the ride on our tour was 86-year-old Commander Fred 'Baldy' Baldwin, Gene's lifelong friend and a decorated United States Navy pilot. From the back of the ten-seater Beechcraft Super King Air 200 twin-propeller plane, Gene yelled: 'It's hot as hell in here!'

'Yeah—I'm trying to fix it now,' I called, twisting and turning the plastic air outlet back and forth with frankly no idea what I was doing. Heated air billowed from the vents on the floor, raising the temperature inside our small metal tube to a sweltering, almost frightening level as we rose in altitude. My shoes and socks came off and I rolled my trousers up and stared down at my swollen feet as my thoughts flitted between the remarkable predicament I found myself in and whether we were all going to die a fiery death.

In his incredible career, Gene flew to the moon not once but twice. He piloted the lunar module of Apollo 10, the mission that ventured just 15 kilometres from the lunar surface and paved the way for the first moon landings a few months after that. He also commanded Apollo 17, the final crewed mission to the moon, during which he spent three full days exploring the lunar surface and travelled more than 35 kilometres across the Taurus-Littrow valley on a rover, studying its geology.

'Erm, it really is getting very hot back here!' I called a second time with typical British understatement to our pilot, who seemed oblivious to the fearsome heat.

'It's OK—I've got the aircon on. It should cool down in a few minutes,' he reassured me, in typical Australian style.

Hot air continued to spew from the floor. The metal strip around the carpet had become too hot to touch. My mind conjured up images of orange flames dancing and crackling beneath the flimsy nylon-carpeted floor and our small plane falling from the sky leaving a trail of smoke like an acrobatic display team.

A few minutes passed and the chatter stopped as we exchanged glances. We'd been in small planes in the outback before and knew this was not normal. Dissent started coming from the passengers, and finally Baldy, a lifelong aviator with almost 4000 flight hours to his name, decided to take charge.

'Young man,' he said to the pilot in his softly lilting western drawl, 'where's your flight manual?' The pilot indicated to his right. With the agility of someone a quarter of his age, Baldy climbed forward and rooted through the glove compartment. Pulling out the large bound folder, he flicked through the laminated pages until he found the instructions for the air-conditioning system. Slowly and deliberately, the experienced pilot located the problem and flicked a few switches. 'That oughta do it,' he said with satisfaction.

For what seemed like an eternity, tepid air continued to fill the cabin. But slowly, relief came. Finally, cool air kissed our faces, which came like the relief of finding a waterhole in the desert. Everyone relaxed a little, and the mood lifted as we began to chatter excitedly above the glimmering azure water of Shark Bay. A lone voice—Gene's—called from the back 'It's too cold!', before he flashed a deliberate sideways smile towards his old friend. In response, Baldy made a winding motion and slowly raised his middle finger like a drawbridge. Looked like we were going to make our final show after all.

As we descended towards Carnarvon, everyone was on a high from the magnificent scenery and more than a little relieved that we hadn't perished. We circled the town once, spotting the dazzling white dishes of the NASA tracking station to the north of the town that would host a VIP reception for us that afternoon, and coasted gently down towards the runway.

On approach everything seemed serene until the moment of landing, when we smacked down with some force. The tiny plane lurched to one side, skidding slightly and bouncing a few times. My stomach danced in my throat as I had my second (I thought) near-death experience of the day. As the pilot regained control, Baldy, now sitting towards the front, glanced back at me with narrowed eyes and mouthed some choice words that I'm guessing he picked up in the air force. The plane shuddered to a halt next to a neat white hangar and the crowds of people lining the airport fence greeted us, smiling and waving, awaiting a glimpse of their famous VIP visitor, the moon man.

As the pilot walked down the aisle, Gene boomed with no hint of sarcasm: 'Great landing, son.'

'Great landing, my ass!' retorted Baldy instinctively and slightly too loudly. The pilot pretended not to hear. I would normally have been embarrassed on his behalf, but to be honest I just wanted to get the hell out of that plane.

Why had it touched down with a bang? As Baldy later explained, our air speed had been too low on approach, so at the critical moment of landing, the gravitational force between Earth and the aeroplane was stronger than the lift we experienced due to the airflow around the wings. This caused the plane to land with a bump.

Gravity doesn't just affect us here on Earth—it also shapes our universe and is the choreographer of many great cosmic dances. It is the invisible force behind spiral galaxies, with giant reservoirs of material contributing to their gravitational bulk.

Our Milky Way contains 50 billion solar masses of gas in the form of stars, and around 10 billion solar masses residing in gas clouds. It also seems to hold a whopping 88 per cent of its bulk in a mysterious material called *dark matter*, which in our galaxy alone exerts a gravitational force in excess of one billion billion times Earth's gravity.

Dark matter is as freakish as it sounds—it's a completely invisible material made from a totally unknown substance. I know that sounds like I've just made it up, but in fact there is ample evidence for the stuff. We believe that dark matter is there because, like a bird flying high above Earth can feel gravity, we can feel its gravity without touching it directly.

That there was matter 'missing' from the universe was first noticed by Swiss astronomer Fritz Zwicky. In the early 1930s he measured the velocities of around 800 galaxies in the Coma cluster, a vast collection of galaxies more than 300 million light years from Earth. He predicted, using an estimate of the mass of the cluster arrived at by counting all the stars and gas, that the average velocity of the galaxies should be in the region of 80 kilometres per second. However, measurements showed that the actual velocity of the galaxies was on average 1000 kilometres per second. This led Zwicky to the conclusion that there was far more unseen material—which was exerting this additional gravitational pull—than visible material within the Coma cluster.

This missing matter doesn't just lie in clusters of galaxies—it also hides within individual galaxies. When an American PhD student called Horace Babcock measured the motions of gas and clusters of stars in the Andromeda spiral galaxy in 1939, he noticed something baffling. For the stars towards the centre of a galaxy, the rotation speeds were pretty much what he expected given the mass that he couldn't see in stars and gas. In the outer regions of the galaxy, however, the orbital speed of stars and gas was more than twice as fast as he expected from just adding up the gravity from all the visible material.

This effect was not a one-off—it was seen again and again, in every spiral galaxy. Astronomers Vera Rubin and Kent Ford were the first to figure out the amazing implications of this observation, which completely changed our understanding of the composition of the universe. To explain the observed rotation, they realised that the 'unseen' mass must be distributed largely towards the outskirts of galaxies. In other words, galaxies are surrounded by a large spherical distribution of material that cannot be seen. This is called a *dark matter halo*, and is closely associated with, but far bigger than, the halo of stars and gas that surrounds all galaxies.

When we stare into the distant universe with our most powerful telescopes, we see stretched and distorted images of distant galaxies. In some cases, we see two or more images of the same galaxy or a crescent or a complete ring. These are images caused by the bending of light by gravity, an effect called *gravitational lensing*.

Gravitationally lensed images of background galaxies are often seen close to foreground clusters of galaxies. By studying the stretching of light from background galaxies, it is possible to calculate the amount and the distribution of mass in the foreground galaxy cluster that is producing the lensing. This reveals that only 20 per cent or so of the mass in a galaxy cluster is made from stars and gas—things we can see. The remaining 80 per cent is thought to be dark matter.

This has striking implications for the composition of the universe. When we add up what the universe is made of, 27 per cent of all the mass and energy is made up of dark matter. The regular stuff that makes up everything we've ever seen, touched, tasted or smelled comprises only 5 per cent. What is the rest?

The answer is that 68 per cent of everything that exists is made up of dark energy—the force that blows our universe up like a balloon. We don't know how, but it seems to drive galaxy clusters further and further apart as the universe grows exponentially bigger. This effect, called *acceleration of the universe's expansion*, was discovered in the 1990s by two teams of astronomers who were looking at the brightness of distant supernovae. They found that distant supernovae were much dimmer and thus further away than they should be if the universe was expanding uniformly or slowing down due to gravity, as some expected. The team leaders were awarded the Nobel prize for their discovery in 2011, at a ceremony hosted by the Swedish royal family in Stockholm.

I take particular joy in this Nobel prize because I've spent many happy hours at the Maipenrai Vineyard and Winery outside Canberra, which is the home of one of the recipients, Brian Schmidt. At the Schmidts' annual vintage I've helped to pick the fat grapes from which he and his family make cool-climate pinot noir. It is a splendid event where a band of astronomers and their families gather every year to bring in the harvest and chew the fat over a meal. Barbecues sizzle and Brian takes charge of a delicious vegetarian paella, cooked outdoors on a gas burner. I enjoy the grape-picking immensely but must admit it has put me off wine a little. You wouldn't believe how many spiders live in bunches of grapes as they grow on the vine. Those grapes end up being tossed into the mashing machine, and presumably everything on the grapes enters the food chain. It seems that wine has a higher protein content than I originally thought!

Unlike the hidden 'dark matter' in your wine, the dark matter in the universe is not just unseen but, we believe, fundamentally unseeable. That's because despite many careful searches, we can't find a scrap of evidence of light either emitted or absorbed by dark matter haloes. Some scientists have proposed that dark matter is made up of compact objects such as black holes or failed stars. Several teams have embarked upon searches for these objects, using the fact that compact gravitational objects should cause bright flashes of light all over the sky as gravity acts as a lens to magnify background stars and galaxies. Since we don't see this, most astronomers are pretty comfortable with the theory that dark matter is made up of an as-yet-undiscovered material whose only interaction with ordinary matter is by gravity.

Whatever form it takes, dark matter is indisputably a unifying bond. It is the hidden glue that binds galaxies in large groups or clusters, coaxing them to dance together in space. But this bond does have its dangerous side. That galaxies live in clusters means they are more likely to interact with one another than if they were spread uniformly through space. When they get too close, they collide.

The night sky is littered with examples of interacting galaxies. The cosmic train wreck called the Antennae is just one spectacular example—a brilliant cauldron of star formation that is the result of two large galaxies undergoing a head-on smash. The Antennae is a classic example of a so-called *starburst galaxy*, host to dazzling swarms of luminous new stars and forming hundreds of times more rapidly than the Milky Way. Whereas our galaxy makes a measly seven new stars on average every year, the Antennae galaxies are producing at least five times that number.

In visible light, starburst galaxies appear as a mangled wreck of spiral arms and two (or sometimes more) galactic cores linked by stretched limbs of gas and stars. Deep images with the Hubble Space Telescope reveal that right now up to a quarter of galaxies are undergoing some sort of major collision.

Many others are undergoing minor interactions. Large spiral galaxies like the Milky Way often attract a swarm of minor or 'dwarf' galaxies that settle into orbits around them. These orbits can be stable, but also random in direction. In the melee of galaxies whizzing in different directions, interactions will inevitably occur.

When the orbits of the minor galaxies intersect the inner regions of the larger galaxy—for example, the disc—significant disruption can result. Unsurprisingly, since the difference in mass between the two is great, the smaller protagonist often comes off worse.

As the two galaxies approach, the gravitational forces on the dwarf galaxy cause it to stretch in the direction of travel. As it begins to interact with the larger galaxy's disc, gas mingles and forms turbulent eddies, like the random blooms in a plume of smoke being blown in the wind. As stars pass close to one another in the collision zone, their regular circular orbits are disrupted and they are sprayed around like shotgun pellets. Computer simulations and observations show that impacts of this sort can warp the shape of spiral arms and make a galaxy's disc thicker. It can also spell the end for the dwarf galaxy as an independent entity. With each passage, orbital energy is sucked from the smaller partner in the cosmic dance until eventually it slows to such a speed that the majority of its gas and stars are absorbed by the disc of the larger galaxy.

The importance of minor mergers in sustaining the lives of larger galaxies is coming sharply into focus. Observations reveal that disturbed and warped spiral galaxies have higher rates of star formation than unblemished spiral galaxies, suggesting that collisions with minor galaxies are crucial in boosting the continued formation of stars. The discs of galaxies hold only enough hydrogen gas to last a few billion years, so just how they will continue to form stars into the future has been a long-running mystery. Somehow, gas from the outside (either from the galactic halo or from other galaxies) will need to be added to the disc's gas reserves to replenish supplies and ensure that a new generation of stars is possible.

It seems that one way this can happen is for collisions with dwarf galaxies to act as a delivery service, bringing fresh gas in from the outside. Without some sort of cold gas courier, star formation is doomed to end—and the galaxy will rapidly become a stellar retirement home. Let's hope for the sake of the Milky Way's survival that this aerial bombardment continues, giving our galaxy a new lease of life for billions of years to come.

Minor collisions with dwarf galaxies might be important in feeding gas to galaxies, but major mergers, involving two or more large galaxies, can be transformational events. A head-on collision between two large galaxies is a high-stakes game of cosmic marbles. But because galaxies are mostly empty space, with stars separated on average by 50 trillion kilometres, the probability of two stars directly colliding is vanishingly low. In the rare case where stars pass within touching distance of one another, they can get caught together in a gravitational spin and flung off into deep space at hundreds of kilometres per second, like a gravitational slingshot.

When galaxies collide it is primarily the *gas* that interacts, since gas particles are much closer together than stars. As two galaxies approach one another, the gravitational forces between them begin to stretch and warp the gas, in some cases unfurling the spiral arms and creating streams or bridges of material between them. With every single atom of gas that interacts with another, the galaxies lose momentum by a process called *dynamical friction*. Gravity pulls the galaxies together and at the same time slows them down. This has the effect of merging the galaxies on timescales of a few billion years. That sounds like a very long time, but it's because galaxies are pretty bulky things and take a long time to slow down.

Gas in the colliding discs is squashed together, leading to the formation of giant clusters of new stars. Billions of stars are typically created in one of these 'starburst' events, including large numbers of very luminous stars between eight and twenty times the mass of our sun, which otherwise form vary rarely under their own steam. These gigantic stars burn brightly, superheating the gas around them to millions of degrees Celsius. Since hot air expands, this generates a *galactic wind*, like the blast from a hot air balloon, that blows outwards from the centre of the galaxy (where the majority of stars are forming). Galactic winds are visible in X-ray images of starburst galaxies as bright regions spreading out from the galactic disc. We believe that these powerful winds are responsible for purging the galaxy of much of its dense cold gas, which is the fuel for future star formation. Hence, this process essentially dooms the galaxy to a dormant future with no new stars.

Starburst galaxies are quite easy to find: they shine like floodlights at infra-red wavelengths. That's because they contain vast quantities of gas and 'dust'—the name astronomers give to chemically enriched clouds of hydrogen-, carbon-, nitrogen-, silicon- and oxygen-based molecules. Molecular clouds are formed when atoms generated inside stars by nuclear fusion are returned to deep space at the end of a star's life. This process can happen gradually, when a sunlike star develops a fast wind that blows the outer layers of the star into interstellar space, or very suddenly, in a supernova explosion.

Either way, the result is a large volume of gaseous material from inside a star spilling into space, where it cools and gathers together to form molecular clouds. When the molecules are gently warmed by nearby stars, they glow infra-red. With the ferocious levels of star birth (and therefore also star death) happening in a starburst galaxy, there is an abundance of molecular clouds and stellar radiation that creates a perfect firestorm of infra-red emission. With infra-red telescopes we see these galaxies shining like beacons across vast distances of the universe.

Early in my research career, a colleague and I embarked on a project to map out hidden regions of massive star formation in starburst galaxies using the Effelsberg radio telescope in Germany. You might think that young stars weighing more than 20,000,000,000,000,000,000,000,000,000,000,000 kilograms and shining at 30,000 degrees Celsius would be easy to spot, but strangely that's not the case. Since they are born deep inside dense, dusty clouds of molecular gas, they are invisible to even the best optical telescopes in the world. But radio waves travel through dense molecular clouds unimpeded, so in our experiment we were going to use *methanol masers* as signposts to these stellar nurseries.

Methanol masers are microwave lasers—beams of radio waves emitting tremendously brightly at a frequency of 6.7 gigahertz. They are generated by gentle heating of molecules of methyl alcohol by infra-red radiation in space, and are found only in the vicinity of very young stars with masses greater than eight times the mass of the sun. That means if we tune a radio telescope to 6.7 gigahertz, we can easily locate and map out all the massive stellar nurseries in the Milky Way.

Methanol masers had never been seen in other galaxies, so we knew it would be a tricky observation to get right. But I'd done the maths, and if the methanol masers in starburst galaxies matched the brightest ones seen in the Milky Way, we should easily be able to detect them if we could use the Effelsberg telescope. After all, with a diameter of 100 metres and a surface area of almost 0.8 hectares, it was the second-largest fully steerable radio telescope in the world. We emailed our idea to the director of the Effelsberg telescope and within a couple of weeks had been awarded time on the magnificent instrument to pursue our scientific goals.

My confidence was high. I had teamed up with an experienced galaxy wrangler—an affable Dutch radio astronomer called Willem Baan, who worked alongside me at the Dwingeloo Radio Observatory in a sleepy village in the Netherlands. Almost at retirement age (although astronomers rarely retire), Willem was a highly experienced radio astronomer with some significant observational successes in the bag. I couldn't have hoped for a better companion for our trip.

We headed off in Willem's car towards the Effelsberg telescope near Bonn. I don't know what scared me more during the three-and-a-half-hour drive: the unconstrained speeds of the German cars overtaking us on the autobahn, or the prospect of two and a half more hours of my colleague's religious folk music. Still, after winding our way through the vineyards of North Rhine-Westphalia and doubtless earning God's blessing with our musical devotion, we arrived safely at the observatory. I was struck by the sharp contrast between the bright sunshine of the surrounding countryside and the cool, dark valley that is home to the telescope. The valley provides at least some protection from radio waves straying from nearby towns and cities that would otherwise drown out our targets—the faint radio signals from the night sky.

Parking up at the small concrete observatory building, I dropped off my bag in my basic quarters and had a look around. The telescope control room looked rather like a bunker but had large windows with a very impressive outlook onto the telescope. Next door was a recreation room with a television. I clicked the green button to see what German TV was like, and to my embarrassment was faced with an 'adult' channel. Scrambling for the remote, I just managed to turn it off before one of the telescope staff wandered in. Needless to say I didn't dare turn the TV on again for the rest of the trip.

Moving on quickly, I entered the heart of an astronomical observatory: its kitchen. The gastronomical standards of observatories vary significantly around the world. Some, like Arecibo in Puerto Rico and the Australia Telescope Compact Array near Narrabri in New South Wales, employ cooks and provide three square meals a day for observers. Many of my observing trips to Narrabri have ended in a soft belly after I have stuffed it full of the daily desserts and freshly baked cakes at 'smoko' (morning tea break). Other telescopes, like the one at Parkes, have a kitchen inside the telescope building itself, since you can't leave the telescope unattended for even a minute while you are observing. Parkes staff lay on oodles of exciting goodies, such as biscuits, nuts, cheese and crackers and even microwave meals. Here at Effelsberg, the cupboards were pretty bare as this was a self-catering joint. I had rather hoped for some local beers like they serve at the Max Planck Institute for Radioastronomy in Bonn and at the Sterrewacht observatory in Dwingeloo, but I was sadly disappointed.

Willem emerged to snap me out of my culinary musings and we agreed to rendezvous back at the control room. He was eager to teach me how to use the telescope systems (they are all different), and he did so before we settled into routine preparations for a long night of observing. We checked our science case and settled on our target galaxies, choosing bright galaxies to calibrate the instrument and making computer files filled with instructions to the telescope in readiness for the night's observations. We started observing before the sun set. That might sound strange, but to radio astronomers day and night have little meaning. They are optical constructs, and radio waves can be seen just as well during the daytime as during night-time. So long as you don't point the telescope too close to the sun, everything is generally OK.

For the first night we stared at the nearby starburst galaxy M82, which has much more active star formation than the Milky Way. If a methanol maser signal were present it would appear as a 'spike' on the computer graph in front of us, but all we saw was static. Glancing at it again and again throughout the night, we eagerly awaited the emergence of a faint signal as the data accumulated over several hours. To check that the instruments were working, we briefly pointed the telescope at a bright maser in our galaxy and watched the peak rise quickly into something that resembled the Matterhorn. The telescope was definitely working, so back we pointed to our starburst galaxy, M82, to keep trying.

I had plenty of time to think on that observing night. As I waited, growing increasingly delirious as one coffee faded into the next and a third cheese toastie seemed like a bridge too far, my mind wandered to the astronomers throughout history who have watched and waited for their astronomical quarry. I imagined the comet hunters, like Caroline Herschel and Charles Messier, trudging back into their houses, chilled to the bone and exhausted, after spending countless hours fruitlessly scanning the skies for a fuzzy ball of light that was missing from the star maps. And I thought back to the ancient Greek, Roman, Mesopotamian and Chinese astronomers tirelessly mapping each star to create charts for navigators and astrological charts to counsel kings. *At least I'm inside in the warm with a hot drink*, I mused.

Over five consecutive nights we continued to search our chosen group of starburst galaxies for massive stellar nurseries, but all we gathered was static. So what did we learn?

Either methanol was not abundant enough in the galaxies we searched, or the signal was simply too faint for us to see. Scientists learn to be resilient in such circumstances. Countless hours of astronomical observations yield null or unremarkable results, but these are not mistakes. Searching and not finding is a result in itself. We learn about nature's limits from this process. What some might call failure is a vital part of the scientific quest for something special that will deliver us new insight into the workings of the universe.

The merging of two galaxy discs sets off a cacophony of star formation that lights up the sky like a sea of fireworks. But that is not the only show put on by a galactic impact. As the two galaxies slow down and settle ever closer and their cores start to merge, the supermassive black holes at their centres can combine. Computer simulations show that in large galaxy collisions, the black holes from each constituent galaxy will eventually merge to form one. This process can unleash some of the most spectacular cosmic forces imaginable.

Astronomers recently observed a starburst galaxy with a pair of supermassive black holes on what seems to be a direct collision course. The system, called NGC 6240,[4] in the constellation of Ophiuchus, is a mangled mess of spiral arms and dust—the result of the collision of two spiral galaxies. Optical, radio and infra-red images show a pair of bright compact objects at the centre, and a picture taken with the Chandra X-ray telescope that orbits Earth proves that they are a pair of supermassive black holes. We have come to this conclusion because high-energy X-rays are found only in these extreme environments. The distance between the two supermassive black holes is around 4000 light years, which is snug

4 NGC stands for *New General Catalogue*, a list of galaxies compiled by John Dreyer in 1888 and subsequently revised.

TOP LEFT: Posing with my BMX aged 12 in our garden in 1992. This back garden was also used as my first astronomical observatory, albeit with no equipment, around the same time. TOP RIGHT: Pointing to Comet Hyakutake from my dark skies spot in dog shit alley, Wethersfield, UK. I lined up the camera on my tripod, set the camera up for a 20-second exposure, during which time I had to stand very still. Not bad for a 16-year-old beginner!

BOTTOM LEFT: With my dad Dave and sister Cassie on the front step of our Wethersfield home, taken approximately 1999. BOTTOM RIGHT: Celebrating England's progress in the 2002 football World Cup with a friend in Newcastle-upon-Tyne. I was a very serious student of astrophysics at the city's university at the time. *Lisa Harvey-Smith*

We can't see our home galaxy because we're inside it, but this is a pretty accurate artist's rendering of the Milky Way, adapted from NASA/ESA images of similar spiral galaxies. *Nick Risinger, NASA/ESA*

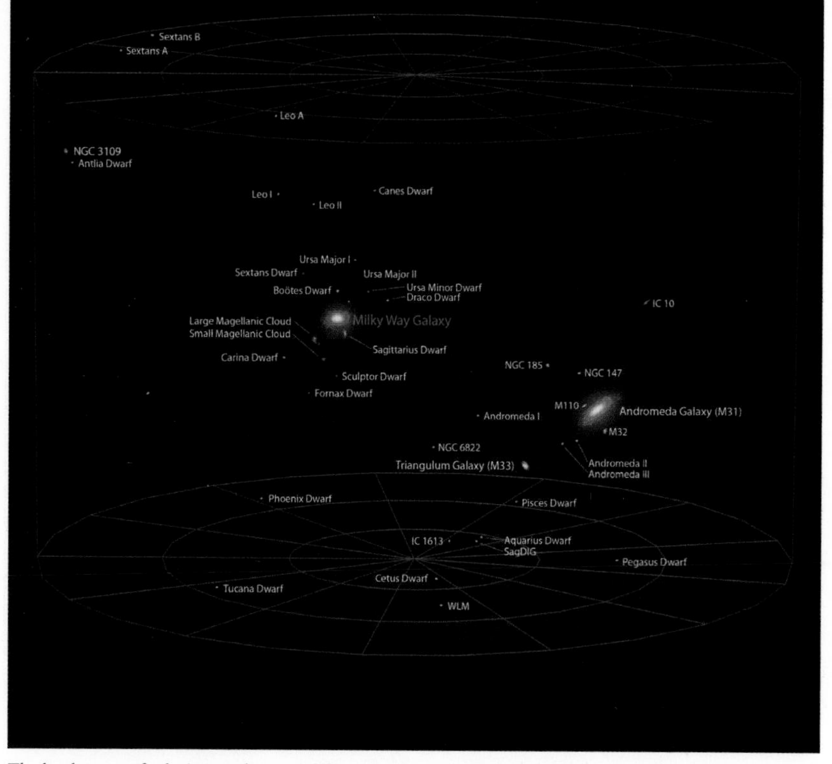

The local group of galaxies, our home and future mosh pit of collisions. *Andrew Z Colvin*

TOP LEFT: On our 2016 Australian tour of 'The Last Man on the Moon', Apollo astronaut Gene Cernan's best friend and decorated US Navy pilot Fred 'Baldy' Baldwin flicks through the manual of our small twin-propellor plane. He was trying to fix the air-conditioning system, which was baking us pretty badly. *Lisa Harvey-Smith* TOP RIGHT: Hard at work picking grapes at Nobel prize-winner Brian Schmidt's place. *Michelle Reid* BOTTOM: An early morning run at Boolardy Homestead near the site of the Square Kilometre Array telescope in Western Australia. This was just before I had a rather exciting run-in with an emu. *Lisa Harvey-Smith*

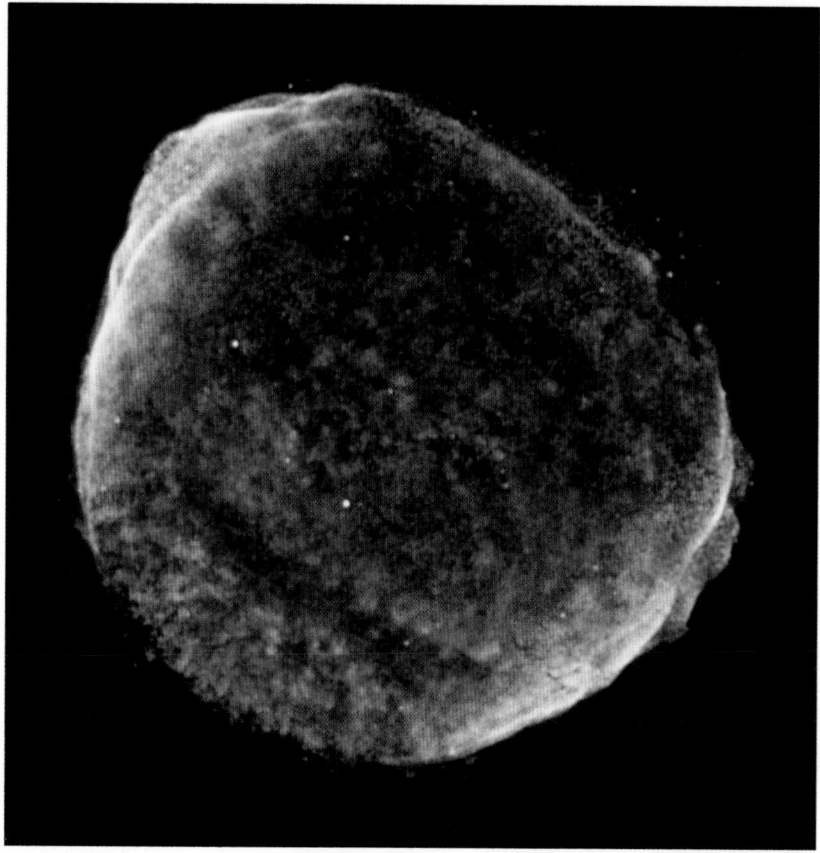

Supernova 1006, which was observed as a dazzling star visible during the daytime by Chinese astronomers when it exploded just over 1,000 years ago. All that is left now is this glowing cloud of gases from the stellar explosion. *NASA, ESA and Zolt Levay (STScI)*

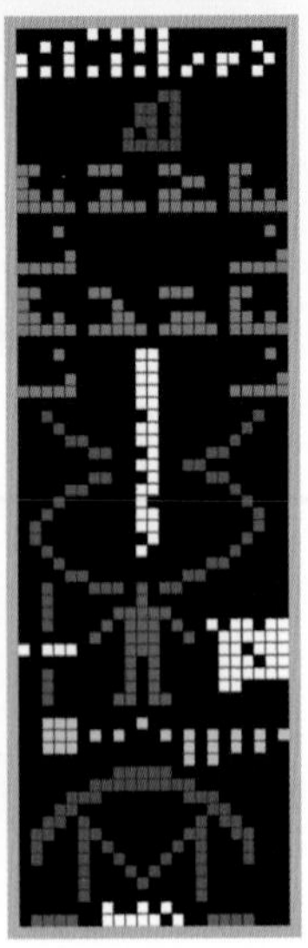

TOP: Willem Baan and me up all night searching in vain for massive stars in nearby galaxies at the Effelsberg radio telescope in Germany. *Lisa Harvey-Smith* FAR LEFT: Trying not to look down as I ascended the Arecibo radio telescope in the wee small hours. *Lisa Harvey-Smith* LEFT: A signal sent into space by humans in 1974, advertising our presence. The message, devised by astronomers Carl Sagan and Frank Drake, was sent towards the globular cluster M13 in binary. It contains mathematics (white), chemicals (green), the structure of DNA (green, white and blue), our body shape (headless, in red), our position in the solar system (yellow) and a representation of the transmitting Arecibo telescope (purple). *This is a pictorial representation by Arne Nordmann (norro)—Own drawing, 2005, CC BY-SA 3.0, https://commons.wikimedia.org/w/index.php?curid=365130*

TOP LEFT: Catching up with paperwork as my observations proceed through the night on the 305-metre diameter Arecibo radio telescope in Puerto Rico. *Lisa Harvey-Smith* TOP RIGHT: My wooden hut at the Arecibo radio telescope, where I had the ★ahem★ incident and locked myself out. *Lisa Harvey-Smith* ABOVE: With Apollo 11 astronaut Buzz Aldrin before the Melbourne show of his live on stage tour 'Buzz Aldrin: Mission to Mars' in 2015. My role was to provide an introduction to the shows.

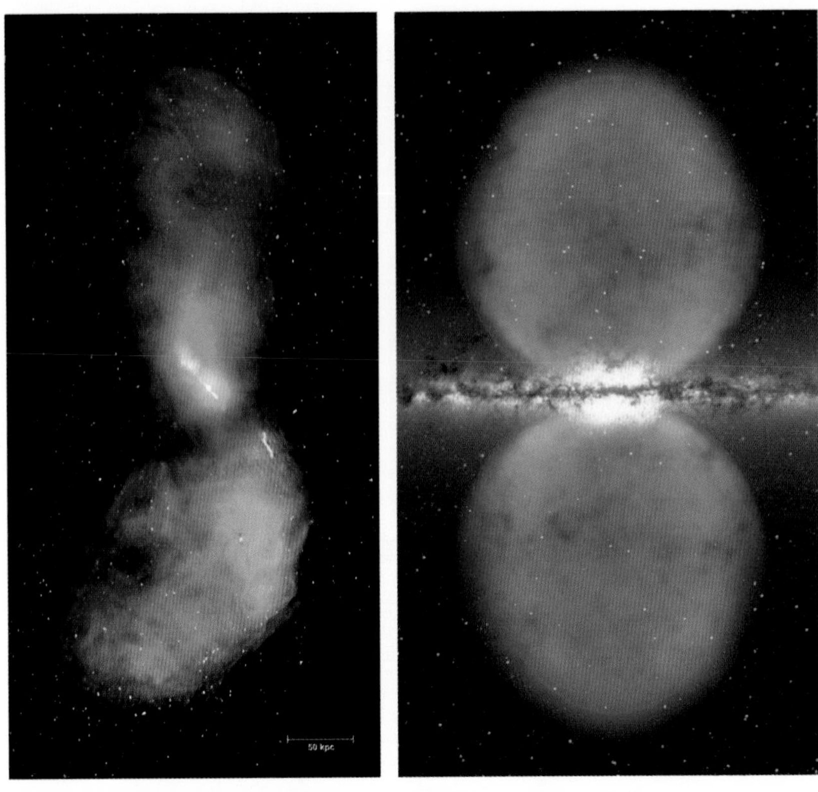

TOP LEFT: The active galaxy Centaurus A. High-energy particles stream millions of light years into space from the heart of the galaxy, where a supermassive black hole is ingesting stars and gas. These particles emit radio waves and are seen by radio telescope as the purple emission in this image. The smallest structure visible in the image is 210 light years across: the scale bar represents about 163,000 light years. *Ilana Feain, Tim Cornwell & Ron Ekers (CSIRO/ATNF). ATCA northern middle lobe pointing courtesy R Morganti (ASTRON), Parkes data courtesy N Junkes (MPIfR)* TOP RIGHT: The Fermi Bubbles (purple) are two huge structures believed to have been expelled by the hot region surrounding the Milky Way's supermassive black hole around 10 million years ago. *NASA's Goddard Space Flight Centre*

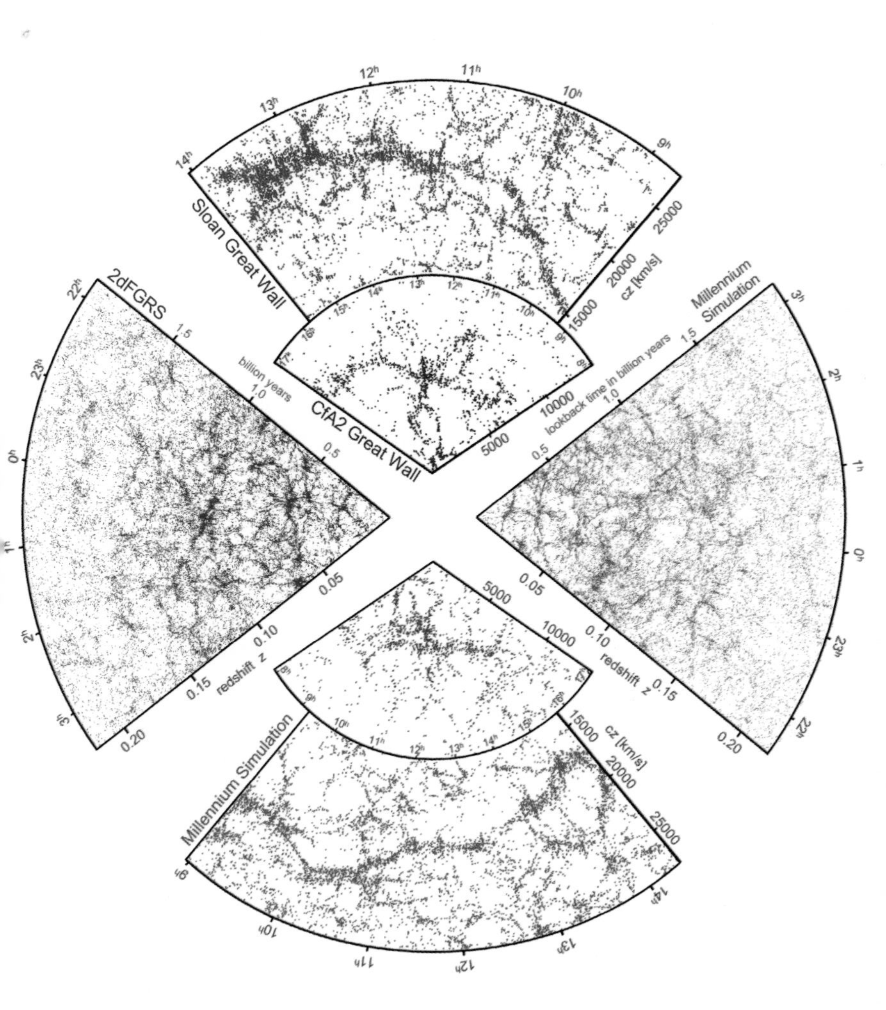

The large-scale structure of the universe—the 'cosmic web', showing observation versus theory. Each tiny point in these images is a galaxy. The distributions of galaxies are weaved into complex structures in space. The top and left-hand quadrants are observations looking in different directions in the sky. The right-hand and bottom quadrants show computer simulations of the large-scale structure of galaxies in the universe. The fact that simulations and observations match quite well shows us that our understanding of the evolution of the universe is pretty good. *Volker Springel, Max-Planck Institute for Astrophysics, Germany*

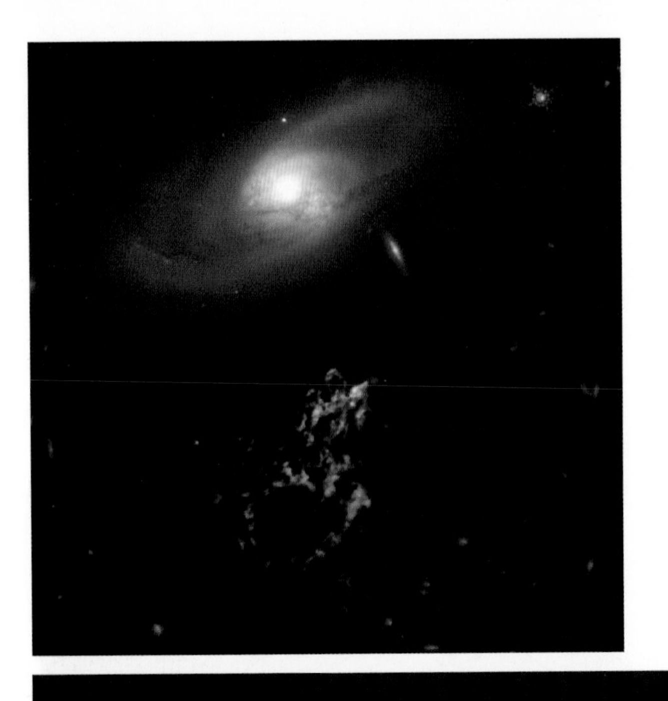

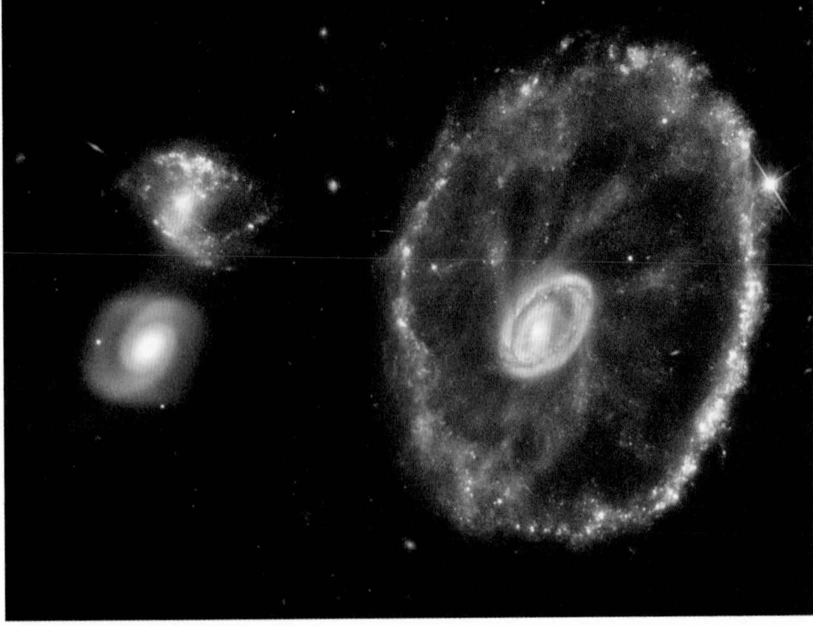

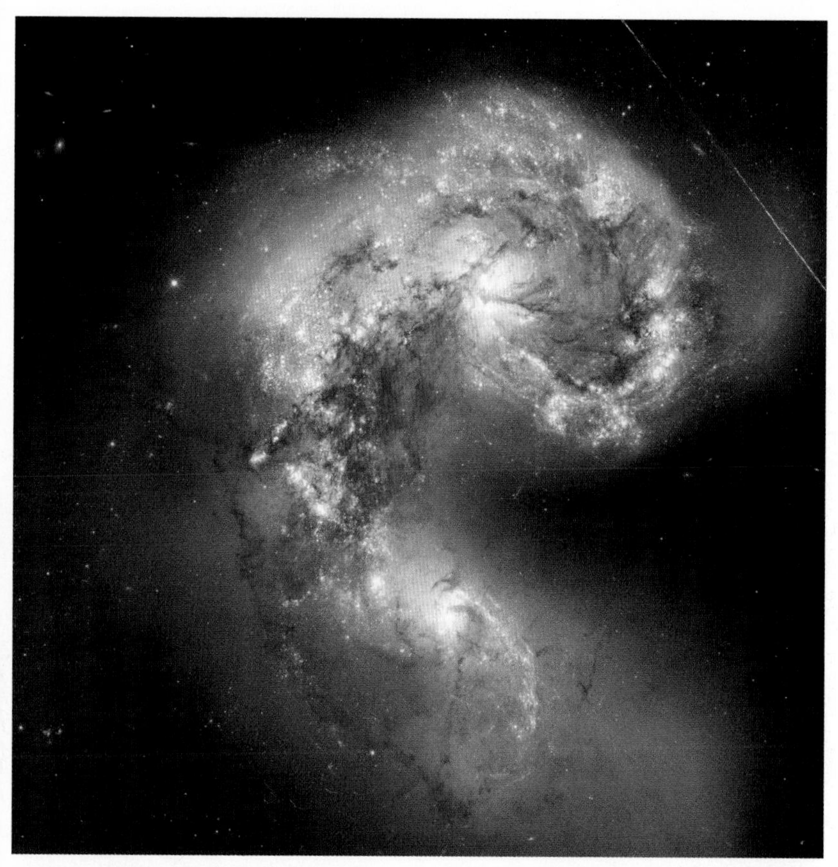

OPPOSITE TOP: The strange object called 'Hanny's Voorwerp', discovered by Dutch school teacher Hanny van Arkel. The green shows an enormous cloud of gas, illuminated by the energy of a supermassive black hole. *X-ray: NASA/CXC/ETH Zurich/L. Sartori et al, Optical: NASA/STScI* OPPOSITE BOTTOM: The Cartwheel irregular galaxy is the complex stucture seen on the right of this image. The galaxy's nucleus is the bright yellowish spiral at the centre and the spokelike structures are composed of stars and gas. The galaxy's weird look was created by a nearly head-on collision with a smaller galaxy around 200 million years ago. *Hubble Space Telescope, NASA/ESA and STScI* ABOVE: The Antennae galaxies, a pair of colliding spiral galaxies ripping each other apart and devouring the gas in a frenzy of star formation. *The Hubble Space Telescope, NASA/ESA*

An illustrated montage of the future of the night sky as the Milky Way and Andromeda merge. Row 1, left: Today the Andromeda Galaxy appears as a faint smudge. Row 1, right: In 2 billion years the disk of the approaching Andromeda Galaxy is noticeably larger. Row 2, left: In 3.75 billion years Andromeda dominates the stars. Row 2, right: In 3.85 billion years with the collision in full swing and gas colliding all over the place, our sky is filled with newly forming stars. Row 3, left: In 3.9 billion years, the glittering night sky of a starburst galaxy. Row 3, right: In 4 billion years both galaxies are stretched by each other's gravity. Row 4, left: In 5.1 billion years the cores of the Milky Way and Andromeda appear as a pair of bright lobes. Row 4, right: In 7 billion years' time we will live in an elliptical galaxy with a bright night sky. That's unless we've been flung out into deep space. *NASA, ESA, Z Levay and R van der Marel (STScI), T Hallas, and A Mellinger*

Light pollution on the Iberian peninsula as seen from the International Space Station. It's a wonder we can see any stars at all. *NASA*

by normal standards but not close enough to bind them by gravity. The future for this pair of black holes is even more intimate, as drag forces cause them to slow down and their orbit to shrink.

Images of another galaxy, NGC 7674, taken with a radio telescope with a formidable zoom lens, show two black holes dancing in a close embrace, with a separation of *less than 1 light year* and an orbital period of 100,000 years. Eventually, this pair will merge as their event horizons touch, and with a magician's flourish they will become one.

Although we've never witnessed it, the coalescence of supermassive black holes is expected to generate strong gravitational waves that will radiate into space at the speed of light. Gravitational waves are produced by any mass that is accelerating, so go ahead—run around in circles and you too can be a source of gravitational waves! Since a supermassive black hole has around a trillion trillion trillion times more mass than you do, the energy released by this process is tremendous. When two black holes were first witnessed merging by the Laser Interferometer Gravitational-Wave Observatory (LIGO) in 2015, more power was radiated by the event than the combined light emitted by all the stars in the universe. Think about that for a minute.

That merger involved a measly sixty-five times the mass of the sun. The energy released by the merger of two supermassive black holes would be tens of thousands of times greater. The tremendous release of energy as gravitational waves suck energy from the orbit of both black holes draws them ever closer to one another. The gravitational waves continue to increase in power and frequency until the critical moment when the event horizons of the two black holes interact. At this point we detect a final 'chirp' of gravitational energy and then nothing. Silence. As soon as the black holes have merged, there is no further emission.

Theoretical astrophysicists predict that the merging of two supermassive black holes will be visible in more than just gravitational waves. The interaction of the two accretion discs—the searing-hot whirlpools of magnetised gas that orbit black holes—will generate an increase in infra-red light and radio, X-ray and gamma-ray emission from the gas during, and for up to thousands of years following, the merger. Whether those radiation signatures actually exist and what they look like remains to be seen—we'll just have to wait until we detect gravitational waves emitted by a coalescing pair of supermassive black holes.

We might have to wait another fifteen years for that. The frequencies of the gravitational waves emitted by such events will be far too high to be detected by LIGO or the Virgo observatory here on Earth. Scientists are planning ahead, though, and are building an Earth-orbiting gravitational wave detector called the Laser Interferometer Space Antenna (LISA), which is due to be launched in 2034. LISA will use lasers bouncing between three orbiting mirrors spaced by 2.5 million kilometres to detect high-frequency gravitational waves. Assuming that the experiment is a success, the discoveries made by LISA will undoubtedly be a wonderful addition to our knowledge in this new field of astrophysics.

Although there is little chance of directly detecting the merging of two supermassive black holes before then, astronomers predict that it will be possible to measure a faint rumble of gravitational waves from all the supermassive black holes that have merged across the universe. This is akin to the background noise in a busy cafe, where you can't hear individual conversations but there is a general hubbub of sound. My colleagues have been searching for this background noise of gravitational waves for more than a decade now, by looking for subtle changes in the regularly flashing radio signals from a type of star called a *pulsar*. As a gravitational wave passes, the regular 'tick-tock' of the pulsar's radio beam will be disrupted slightly, betraying the presence of a background of ripples in space and time triggered by the merging of supermassive black holes all over the universe. It is hoped that by studying lots of pulsars in different parts of the sky, they will betray this background noise of gravitational waves by their subtle motions, like boats bobbing up and down on the sea.

Only time will tell who will be first to detect gravitational wave signals from the merging of supermassive black holes, which mark the spectacular crescendo of a billions-of-years-long process. Will it be through observations of pulsars, or via a direct measurement of a single merging event by LISA? Whoever gets there first, the discovery will open up thrilling new possibilities for studying these unique events.

5

TRAVELLING THROUGH TIME

The problem with our current understanding of galaxy formation is that humans don't stick around on Earth for long enough. We can't sit and watch a galaxy change or merge with its neighbours within a human lifespan. The only practical approach for astronomers to observe these processes is to take snapshots of millions of galaxies at different stages in their lives and get clues towards building an overall picture of how they grow. Fortunately, we observational astronomers are capable time travellers, at least in the retrospective sense. Observing the sky with powerful telescopes enables us to thumb through the pages of the cosmic history books.

This is made possible due to the universal speed limit—the fact that light travels at a finite speed (299,792 kilometres per second). The most distant galaxy ever photographed is around 13.4 billion light years away from Earth. That means the light entering our telescopes left the galaxy as it was forming from the primordial gas just after the Big Bang, which happened about 13.8 billion years ago. This trick of the light makes it possible for us to see faraway galaxies as they were many billions of years ago. In this way, our telescopes are time machines.

We can't easily observe how galaxies formed, since the light arriving at Earth now from those galaxies is extremely faint. Even so, evidence gathered from photographing galaxies both near and far gives us clues to how they have formed and evolved. Theories are compared with models created in supercomputers to try to figure out how these complex and dynamic systems have evolved through the first thirteen or so billion years of their lives.

How astronomers develop and test such models is a painstaking process. We start with an image of the universe a mere 380,000 years after the Big Bang. It's called the *cosmic microwave background* and it tells the story of a universe with no stars and no galaxies, only gas. The 380,000-year-old universe was incredibly smooth—it shows only a one-part-in-100,000 variation in density throughout the entire cosmos. It is thought that the tiny lumps and bumps in its structure came about from random quantum mechanical fluctuations in the first few seconds of time, where pairs of subatomic particles popped into existence for a fraction of a second before disappearing again in a puff of pure energy. These very primitive structures were amplified by the rapid expansion of space-time, planting the seeds for the entire structure of the universe we see today.

Astronomical computing specialists develop computer models to show us how such a universe would evolve under the action of gravity. How does creating a computer simulation work in practice?

Starting with a lumpy universe filled with dark matter and gas, the computational astronomer will add physical laws such as gravity, pressure and dark energy and press 'Go'. The computer fast-forwards through billions of years, creating a movie of what happens to each and every particle under the influence of the laws of physics. What emerges is a universe full of spiral, elliptical and dwarf galaxies huddled into clusters that is similar to what we observe today. Depending on the initial conditions and the physical laws that are entered into the computer simulation, we can compare these model universes with the real thing and see how well they match. Models can be tweaked and changed as our understanding evolves.

Ultimately, the closer the match, the closer we are to understanding how the cosmos was moulded into the form we experience today.

It sounds simple, but it is actually a complex task to compute the action of all the known physical forces (gravity, magnetism, nuclear forces) acting on trillions upon trillions of particles. As scientists we are often limited not by our imaginations but by the computing power that currently exists to generate our theoretical universes. Despite these limitations, we are slowly making progress.

Our current understanding is that the first seeds of the galaxy formation process were planted around thirteen billion years ago when the universe was a featureless wasteland. Tiny variations in its density from one place to the next became more pronounced as clumps of dark matter accumulated under the action of gravity. Dark matter can collapse relatively easily under the influence of gravity, whereas ordinary matter is subject to other forces such as gas pressure, which means that it accumulates somewhat more slowly. Nonetheless, pockets of gas slowly collapsed over the first few hundred million years to form the first stars.

Simulations suggest that these stars may have been highly unusual as far as today's stars go, being incredibly hot and roughly a million times more luminous than the sun, and holding tens of times more gas than the biggest stars that exist today. We think this first generation of stars grew so big because of their incredibly pure chemical make-up of hydrogen and small amounts of helium, and the warm environment in which they formed. Their gigantic size meant that they would have burned through their nuclear fuel in only a few million years, thousands of times faster than our sun. Once the fuel was spent, their vast mass created a catastrophic gravitational collapse that sparked a supernova and expelled the newly formed elements—carbon, nitrogen, oxygen and others—into the galaxy. As the sun was formed along with our solar system, these atoms became the raw materials that make up planet Earth and every cell of you and me.

Around the same time the first stars were forming, the first galaxies were taking shape. Dark matter collapsed, bringing gas with it to form large, near-spherical accumulations of material. Within these collections of primordial matter, random motions in one or the other direction became stronger and the gas slowly flattened into a disc. Spherical condensations on a small scale formed clusters of stars.

We know that black holes with masses in excess of a billion suns existed in the early universe, because we have measured them in dozens of distant galaxies whose light is only just reaching us from the first billion years of the universe's history. This means that black holes formed alongside the formation of galaxies, from the wholesale gravitational collapse of the clouds that produced the galaxies. Just as a stone sinks to the bottom of a pond, the material in a galaxy slowly accumulates at the centre and collapses to form a black hole.

But evidence from a handful of very early galaxies, only 870 million years after the Big Bang, suggests that the black holes might even have formed first, with the rest of the galaxy accumulating later. That's because in those early days the black holes occupied a far greater fraction of the galaxies' mass than of the galaxies we see today. Computer simulations support that picture, showing that black holes with tens of thousands of times the mass of the sun could potentially have taken shape from turbulent gas even before individual galaxies began to form. We're not certain which came first—the galaxies or the black holes. It is possible that the supermassive black holes we see today originated from a combination of these formation processes.

Galaxies did not form in isolation—in fact, most formed in clusters. Neighbouring galaxies frequently interacted in those early years of the universe, triggering rapid star formation, particularly towards the centre of growing galaxies. Based on our computer predictions, the first billion years of each galaxy's life were highlighted by a spectacular burst of star formation that was fifty times more rapid than the rate at which stars are created today and led to the relatively high abundances of heavier elements (which are synthesised in stars) in the central bulges of galaxies.

As time went by, large galaxies were subject to bombardment by minor galaxies that often became subsumed, and whose stars and gas accumulated in larger pools. Such interactions heated the discs of the larger galaxy, and lone stars and globular clusters were hurled off into the halo. In those turbulent early days, galaxies also ran the risk of major mergers, during which catastrophic tidal disruption of the disc can lead to the formation of an elliptical galaxy.

Elliptical galaxies are almost spherical in shape and their stars and gas move in random directions. They are much more common towards the centre of galaxy clusters, in environments where many galaxies formed within a close-knit community. Astronomers have noticed a strong relationship between the mass of elliptical galaxies and the mass of the supermassive black holes at their centres. All these strands of evidence point towards elliptical galaxies being the result of catastrophic mergers between two large galaxies, often within the environment of a galaxy cluster.

As galaxies collide, their shapes are often disrupted in dramatic ways, creating the irregular galaxies that we see in the night sky. In many cases a burst of star formation is triggered, and with the onset of galactic winds, vast quantities of gas are swept out into deep space, thereby quenching the galaxies' potential to form new stars in the future. This is a probable reason why most elliptical galaxies have little interstellar gas and an almost complete lack of star-forming activity.

Major galaxy mergers often precipitate the awakening of supermassive black holes. Previously stable orbits of stars and gas clouds are disrupted, which can send them perilously close to the black holes at the centres of the galaxies and lead to some pretty ugly cannibalism. As stars and gas start their death spiral towards a supermassive black hole, they form a swirling disc of accreting gas that heats to ferocious temperatures due to the frictional forces. Try rubbing your hands together for fifteen seconds—how do your palms feel? As gas rubs together in the accretion disc of a supermassive black hole, its temperature rises to over a million degrees Celsius, causing it to sing out in light, heat and even X-rays. We can see this emission from the centres of active galaxies right across the universe.

In some cases, the centres of two galaxies score a direct hit and vast quantities of gas stream quickly onto the accretion disc. As the new material streams in, it interacts with the gas in the accretion disc and a powerful beam of searing-hot gas and ionised particles is ejected violently into deep space along with some of the galaxies' star-forming gas. This and galactic winds set up by starburst activity are two processes by which star formation is quenched in merged galaxies, leaving behind old, red, dead elliptical galaxies.

As part of my work in testing a brand-new radio telescope array in Western Australia, I recently turned my attention to IRAS 20100-4156, a system of three interacting galaxies lying about 1.8 billion light years from Earth. It is one of the most luminous starbursts known and contains a black hole at the centre of a mangled galactic wreck. Using the Australian Square Kilometre Array Pathfinder (ASKAP) telescope along with an array of six radio telescopes overlooked by Mount Kaputar in north-western New South Wales, I tuned in to the screams of hydroxyl gas clouds as they orbited at breakneck speed (600 kilometres every second) around the centre of the galaxy.

My colleagues and I used the Doppler effect to convert the frequency shift in the radio waves to an orbital speed, and with a bit of high school physics we used our measurements of this gas to estimate the mass that lies at the centre of the galaxy. The answer? A phenomenal three billion times the mass of the sun, making it one of the largest supermassive black holes ever found. By comparison, the black hole at the centre of the Milky Way is only four million suns—that's 1000 times smaller. This discovery, although in line with our theory of how black holes merge, was nonetheless flabbergasting!

What's really interesting about this is that we can watch and precisely measure the motions of gas in a galaxy collision that happened 1.8 billion years ago, before the evolution of complex life, when Earth was inhabited by algae and multicellular creatures.

Not only that, but we can weigh the black hole at the centre of the collision and show that it has a far greater mass than that of a black hole in a typical spiral galaxy. A mouthwatering prospect for the future is to measure the masses of black holes in a large variety of environments at different ages in the universe's history, with bigger and more powerful telescopes. Only when we have observed a representative sample of galaxy mergers throughout history will we be able to build a complete picture of how our universe has evolved to be the complex and diverse place it is today.

The picture of galaxy evolution presented here is not the whole story, and our understanding of these processes is not complete, but a new generation of sophisticated modern telescopes with increasingly clever cameras and spectral analysis machines are giving us a clearer view of distant galaxies that emitted their light billions of years ago. The next generation of telescopes, such as the James Webb Space Telescope (successor to the Hubble Space Telescope) and the Square Kilometre Array (SKA) radio telescope, will boost our ability to study the evolution of galaxies.

Even now, the pathfinder telescopes built in preparation for the SKA are opening a viewing portal to the universe as it was during galaxies' early formative years. Using these revolutionary radio telescopes, we plan to study hundreds of millions of galaxies that lie up to 13 billion light years away, to better understand the connections between the shapes of galaxies and their ages and environments. By figuring out the average bulk of supermassive black holes twelve, ten, eight, five, two billion years ago we can hope to test our ideas of how galaxy mergers have affected their evolution throughout the entire history of the universe.

One ambitious project aims to detect signals from hydrogen gas as it was slowly destroyed by searing-hot radiation from massive stars and their supernovae, by tuning in to long-wave signals from the very first stars that existed less than one billion years after the Big Bang. Imagine being the first person to make an image of the imprints of the very first galaxies, a blueprint for our entire universe as we know it today. Even the most battle-hardened astrophysicist will surely have a tear in their eye as they comprehend the magnitude of what they have uncovered. Once we have such an image in our possession, we will be able to thoroughly test our theories of the earliest stages of galaxy evolution.

The SKA will comprise around 130,000 radio detectors in a remote region of Western Australia and 200 radio dishes in the Karoo region of South Africa. That's just phase one of the project: the overall plans are even grander but may take decades to realise. The telescopes at each site have a distinct job and have been designed to tackle a range of complementary scientific questions. The detectors in Australia will work at low frequencies, tuning in to (among other things) the low rumbles of star and galaxy formation from the early stages of the universe. The array of 200 dishes will tune in to hydrogen gas in galaxies from the more developed parts of the cosmos, measuring how stars are formed and how galaxies evolve. A remote location is essential for a highly sensitive radio telescope, since towns and cities are awash with radio waves transmitted by mobile phones, televisions, cars and computers.

I have been heavily involved in the ASKAP project for the past ten years, which is developing technologies on the path to fulfilling our ambition of hosting the SKA. This work has taken me on countless trips to the remote Murchison region of Western Australia—an arid inland cattle-grazing region with Martian-red soil that is blessed by the rich cultural history of the Wajarri Yamatji people, the traditional owners of the land on which the telescope is built. The Murchison Shire is often called 'the shire with no town' as it has a population of only 114 people in an area covering 49,500 square kilometres—that's the same size as the nation of Slovakia. With a population density of one person to every 2000 square kilometres, it is perfect for radio astronomy.

Access to the observatory is strictly controlled—uninvited visitors are not allowed—and involves a journey of two full days from the major cities of Australia's east coast. Travel is generally in groups of two to four people, both for safety and to share the driving load. Before getting behind the wheel for my first trip to the observatory, I had to pass a multi-day outback four-wheel drive certification course that included advanced driver training and learning to change a wheel on my own in rural conditions, and how to de-bog and tow from deep sand. I've got to say it was by far the most interesting corporate training course I've ever done!

Trips to the Murchison usually follow the same formula, and this two-week-long stint—to test a new computer system with a subset of six of our thirty-six telescopes—was no different. Two colleagues and I hopped on a four-and-a-half-hour morning flight from Sydney to Perth, which is always uneventful apart from the crosswind landings that sometimes make the approach 'exciting'. We waited in the crowded Perth airport lounge for a few hours with 200 of our best friends—it was during the mining boom and everyone except us seemed to be wearing fluorescent mining attire. After a long wait and some predictable food, we hopped on the one-hour afternoon flight to the sleepy coastal town of Geraldton, which enjoys perfect sunset views over the Indian Ocean. Once at Geraldton airport we hired a 'mine-ready' four-wheel drive and scooted down to the supermarket to get supplies. On my first trip, I stocked up with muesli bars, fresh juices, nuts and other healthy snacks. I needn't have bothered, though—the food at the ASKAP accommodation is among the best I've ever eaten. In Geraldton, a quiet dinner and an early night are always on the cards. We usually stay at a motel on the edge of town with double glazing and blackout curtains to shut out the noise and light from trucks thundering up and down the nearby North West Coastal Highway.

The following morning we were up early to conduct safety checks on the vehicle and load it for the journey with food, water, fuel and a GPS tracking device in case we got into trouble. We took satellite phones with us, since mobile phone reception is completely absent in the Murchison region—again, making it perfect for radio astronomy. After breakfast and coffee in a cute little cafe overlooking the ocean, home to an adorable puppy to whom we reluctantly said goodbye, we bundled into the four-wheel drive. About an hour of driving through a landscape of rolling green hills brought us to a small inland community called Mullewa. This is the last place to have mobile reception, a shop and bitumen roads. We stopped to take advantage of the public toilet, aka 'the bathroom at the end of the universe'. After sending last-minute text messages and emails, we swapped drivers (my shift) and trundled out of town. A left turn onto the Beringarra-to-Pindar dirt road marked the start of our three-hour adventure.

Grain silos and golden fields of wheat quickly turned to red dirt, rocky outcrops and low scrubby trees as we entered the outback. The flora and fauna changed, too. (If you're lucky enough to travel in wildflower season, the region is sprayed liberally with brilliant pink and white crops of tiny flowers.) I remained ready to slam on the anchors at any time: wedge-tailed eagles are commonly seen swooping down to feed off marsupial roadkill; monitor lizards (bungarra) up to 1.5 metres long lumber casually along the road; and enormous outback cattle dart without warning across the dusty carriageway. In summer, dust devils (called 'willy-willies' by the locals) dance around the parched terrain. At any time of year, spectacular thunderstorms can turn the perfect blue sky into a deluge and a flash-flooded road within minutes.

There are not many human-made landmarks along the way. After two hours or so I perked up as we passed the alfresco roadside living room. It started as an old armchair and a TV, and over the years has grown to host an assortment of other quirky objects. Household rubbish has never been so artistic. Later, another landmark was reached as we crossed the Murchison River on a simple wooden causeway, stopping briefly to enjoy the wide trickle of water that gives welcome psychological relief from the heat and dryness of the atmosphere.

Just as the journey seemed it would never end, the dusty orange road hit a small rise, and an unusual rocky outcrop enveloped the road on both sides. As we summited the pass, a small cluster of buildings and a wind turbine rose from the next plain, signalling the first human settlement we had seen for several hours. This is the Pia (rhymes with shire) Wajarri community—a cluster of a dozen or so houses, a school and a community centre. During my years visiting the region, I have spent many hours with the students at Pia Wajarri in classroom activities, giving tours of the telescopes and helping ensure that the kids are able to share in the globally significant astronomical project taking shape in their backyards.

From Pia, our drive continued for another 22 kilometres or so before a hand-painted wooden sign announced that we had finally reached our home base for the trip, Boolardy Homestead. It is a welcome stop, and somewhere to sleep and refuel before the final push to the observatory in the morning. Boolardy was established in about 1876 as a sheep farm, and is now a cattle-mustering station. For many years the old stone cottage at the centre of the site was home to a single family, but today it is a hub of activity, housing engineering and construction workers and occasionally scientific people like myself. The once-sleepy cattle station is home to dozens of workers who are building and commissioning the telescopes.

Staying here is a real outback experience. Staff and visitors sleep in 'dongas'—demountable sleeping accommodation with a small ensuite bathroom each. Eating times are closely regimented, but experiencing the home-cooked meals prepared by the host family has to be the highlight of any trip to Boolardy. Undoubtedly the most enjoyable times of the day are mealtimes—from the full English breakfast I once had that was prepared by a backpacker from Manchester who provided welcome banter in the middle of the outback, to the delicious vegetarian quiche with crumbly, buttery pastry that was caringly made by the host cook.

Early mornings are my only opportunity to get any exercise on these trips. Breakfast starts at 7 a.m. sharp and we need to leave for the observatory at 7.45 a.m., so I set my alarm for 6 a.m., when (thankfully three hours jet-lagged in the right direction from Sydney)

I drag myself up and out to enjoy a morning run. Running is an activity I dearly love, although I've never been overly enamoured with the early morning part. I have been known to go a little over the top and compete in twelve-hour, 100-kilometre, 24-hour and six-day races, so getting up early has been unavoidable over the years.

Wherever I travel in the world, I love exploring on foot. Running connects me with the real sights, sounds and smells of a place. Whether it's a crisp trot through the forests of the Netherlands, a jog through paddy fields in Bali dodging rabid dogs and monkeys, or a noisy and sweaty dash through the mad traffic of Kuala Lumpur, an early morning run gives me the freedom and ability to explore that I just can't get with a taxi ride.

But I know that the Australian outback is different from all these places. In this harsh and alien environment it is easy to lose your bearings. To be safe, I decided that the path less travelled was not a good idea that day. Resigned to padding the well-worn routes, I took off in shorts and a T-shirt down the corrugated access track from Bololardy Station towards the main road.

Two horses grazed by the airstrip, 'Hello!' I boomed cheerily, and received a curious flick of the ears in return. I'll let you into a little secret: I often talk to animals on my runs. A wallaby surprised me once as I turned a corner on a training run, and instinctively I said, 'Hello, gorgeous' to the magnificent little creature whose doe-eyed presence had interrupted my stride. Unfortunately, I was running with my friend Dr Jess Baker, a psychologist and world-class ultra-marathon runner. I always worry about what I say around psychologists in case they are analysing me. Luckily I didn't have to worry about that—in her typical enthusiastic style, Jess burst out: 'Ohmigosh! I thought it was just me who talked to animals!' So I'm not the only one, then.

After about 800 metres I reached the end of the driveway to Boolardy Station and hit the softer sand of the main road. My mouth was already becoming dry. *Right or left?* I mused, knowing full well that the road was barren and featureless for at least 45 kilometres in both directions. I decided to go left and quickly my mind wandered into one of my regular fantasies of running solo from here to the coast. Padding steadily down the road for a few minutes, more alert than usual as I took in the strange sights, I had a sip of water from my hydration backpack. Even at 6.30 a.m. I needed it.

The solitude of that road gave me just the right balance between alertness and fear. My mind flip-flopped between hoping I'd see an interesting outback animal and that I wouldn't run into a hungry dingo pack, an aggressive snake or a human with bad intentions (probably the most scary). That particular thought prompted me to decide upon a turnaround point. Runners generally need to pick a physical landmark at which to turn around—you'll never see us about-face in the middle of a street, or halfway between two lampposts. In my current environment there were no features to placate this particular psychological quirk, so in the end I picked a random bush up ahead and turned around there. Weirdo.

As I trotted back along the sandy road, an emu darted out in front of me and started to run in the same direction. My feathered friend was almost 2 metres tall and had an alarmingly high centre of gravity. My heart raced, partly with the thrill of the chase but also with the sudden injection of adrenaline. Wild animals are not known for their fondness for human proximity. Nonetheless, this adult emu seemed reasonably comfortable that this 168-centimetre, 58-kilogram human didn't pose a threat. It continued to weave erratically up the road about 20 metres ahead of me. As quickly as it had appeared, the bird careered off the road again, disappearing into the bush and leaving me bent over, hands on knees and laughing out loud at my encounter with a modern-day dinosaur. Meep-meep!

Tired from the excitement, I shuffled home more slowly than usual. Alas, there was no time to luxuriate in the shower, and breakfast was rushed. Fortunately, grooming loses its importance in the outback. The routine of work kicked in and the car was quickly loaded, ready for the 45-minute drive up to the observatory. After thirty minutes we passed the 'Welcome to the Murchison

Radio-astronomy Observatory' sign, which reminds drivers to turn off their radio to avoid causing interference with the telescopes. Another ten minutes and we took a sharp left onto the observatory access track. We unlocked the security gates and entered the site.

To our left we finally saw what we were there for: a three-storey-high dish-shaped metal antenna rising like a giant white mushroom from the desert. This was the south-eastern edge of the ASKAP telescope. Driving further into the observatory site across a series of drainage rivulets, more white antennas became apparent as we reached the core of the 36-antenna array that spreads for 6 kilometres across this arid landscape. Further along the observatory road we saw the core of the Murchison Widefield Array telescope, a vast army of knee-high metal radio receivers that look like futuristic robot spiders. Here, an iron statue of a Wajarri man points to the rocky outcrop called Diggidumble that sits proudly at the centre of the observatory site.

I grinned widely as I took in the sight of two of the world's greatest scientific experiments taking shape in this remote dusty region of Australia, poised to study the evolution of galaxies with greater speed and sensitivity than ever before. It seems an unlikely technological hub, but the ancient site is at the cutting edge of science and technology. With these telescopes, we will conduct surveys of the entire sky, photographing and classifying tens of millions of radio galaxies, most of which have never been seen before. Our images will be compared with those from infra-red and X-ray telescopes to confirm whether galaxies are actively forming stars, merging, or sucking material into their black holes. Using data from cutting-edge optical telescopes, we will confirm the distances from Earth to each and every galaxy we find, thus producing timestamps for their ages. All that information will be used to build up a clearer picture of the origin and evolution of galaxies and the role of interactions and mergers.

And my mind flashes to ten years hence, when this will become one of the global epicentres of astronomy as it grows to host the

130,000 antennas of the SKA. Just as archaeologists reconstruct the history of ancient cultures from pieces of broken pottery, astronomers will piece together the lives of galaxies that exist for tens of billions of years from the faint glow they emit into deep space. Who knows what riches we will discover?

6

CRADLE OF LIFE

The Milky Way is our lifeline—the parent galaxy that created us, the cradle of our solar system and home to all the chemical elements required to seed life on Earth. If we could write a biography of the Milky Way, what would it say? Has it fought great battles to reach the state it is in today? Or has it had a peaceful life, slowly growing and nurturing our solar system and the precious life it holds?

Piecing together the Milky Way's history is not easy. After all, we live inside it and therefore can't see directly into its distant past. What we *can* do is gather clues to that history from the chemical make-up, distribution and ages of the stars. Like archaeologists uncovering layers of historical artefacts, astronomers can gather evidence from our night sky to piece together the Milky Way's past and help us to understand how our celestial home came to be.

Our galaxy's age is estimated at 13.6 billion years—almost as old as the universe itself. We age a galaxy by looking at its oldest stars. In the Milky Way, these are located in globular clusters, which surround us in the galactic halo. Stars are aged according to the amount of heavier elements in them, since heavy elements are gradually built up in a galaxy over time as stars live and die. According to this chemical fingerprinting, some of the stars in the Milky Way's globular clusters are around 13.5 billion years old.

Of course, there are many stars in our galaxy that are only just forming now. The raw material that makes up new stars—that is, the interstellar gas from which they are moulded—has been chemically enriched by the atoms fused in the cores of successive generations of stars, their collisions and supernovae. Consequently, newer stars contain complex atoms such as gold, silver and iron. We can measure the abundance of these chemical elements in stars by attaching an instrument called a *spectrograph* to our telescopes. As the spectrograph spreads the starlight into a rainbow, chemicals are seen as dark lines at particular wavelengths. The wavelengths are created when atoms absorb specific colours of the intense light from the star's interior.

Like all spiral galaxies, the Milky Way was formed from the gravitational collapse of a vast region of dark matter and gas. Early gas motions in our galaxy would have been turbulent and chaotic, with many smaller galaxies colliding and interacting with it. But as the number of collisions reduced and the gas settled down, circular motions became more pronounced as the galaxy became smaller and denser and the majority of the gas settled into a single flat layer. Spinning gas is subject to the same physical forces that tried to fling you off a spinning merry-go-round when you were a child. The gas is thereby flattened and squeezed into a thin disc.

The origin of the Milky Way's spiral arms is less well understood. Computer models of disc galaxies show that a density wave set up by an interaction with a satellite galaxy could form a compression in the disc, which then ripples through the disc as a density wave. Squashed or clumped gas is more likely to form stars, so the denser regions give rise to bright ridges of star formation. Since the outer regions of the disc have further to travel than the middle, this quickly sets up a spiral shape in the density wave. This model doesn't explain all the possible shapes and configurations of spiral galaxies, but it comes close to explaining how spiral arms were generated in galaxies like the Milky Way.

The Milky Way actually has two discs. A thick disc of around 3000 light years from top to bottom contains older stars that were created more than twelve billion years ago from the chemically pure mix of hydrogen and helium gas that was available at that time. A much brighter, thinner disc (in which we live) is the hub of activity today, with 95 per cent of all stars residing in it and all the current star formation happening here. Computer simulations show that the thick disc was almost certainly created by interactions between minor galaxies and the Milky Way, which would have been very common in the past. Those interactions heated the existing thin disc of the time, with each event creating a vertical 'step' in the disc's thickness. This is useful, because like the geological history in sedimentary rocks, it allows us to read the lines in the disc and piece together our galaxy's history of major disruptions.

The halo of the Milky Way is around 600,000 light years across and contains 150 globular clusters and at least one billion isolated stars. The gas in the halo is incredibly hot, between 100,000 and 1 million degrees Celsius, which is thought to be due to interaction with a shock wave as it fell from intergalactic space into the vicinity of the Milky Way. Superheated gas remains hot almost indefinitely in such rarefied conditions, due to low levels of interactions between neighbouring particles.

A key question about the halo is whether it formed at the same time as the disc, by freefall gravitational collapse, or whether it slowly built up from external material falling towards the galaxy over time. Of this we cannot be entirely certain. But we can get clues from the composition of the galactic halo, which is still forming and changing even today, more than twelve billion years into the Milky Way's existence. One clue to the history of the galactic halo is that globular clusters towards the inner region of the halo are generally two or three billion years older than those towards the outskirts. Presumably, this means that the globular clusters were not all formed when the Milky Way formed, and that their synthesis was somewhat more sporadic. Astronomers have also studied the age distribution of lone stars in the galactic halo. It shows a similar pattern, with a central concentration of very old stars (11.5–12.5 billion years old) and then a region of slightly younger stars (eleven billion years old) further from the Milky Way's centre. Several galactic streams are visible in these maps, comprising dense lanes of stars of the same age and chemical make-up that presumably originate from a common satellite galaxy.

All of this evidence supports the view that the halo of our galaxy has generally condensed slowly over billions of years, but that this process has been punctuated by several impacts from satellite galaxies, which are gradually becoming incorporated into our galaxy's halo. The prevailing view among most astronomers is that interactions of dwarf galaxies with the Milky Way were at least in part responsible for the high-velocity stars and globular clusters in the halo. Since these events would cause heating in the thin disc (this is not seen—our thin disc is cool and stable), we think that the majority of the interactions that built our galactic halo occurred more than twelve billion years ago.

That the Milky Way is a spiral galaxy tells us that it has probably not undergone a large-scale merger with another giant spiral galaxy in the past. As we have already seen, when a major collision like this happens, the disc and spiral arms are severely disrupted and an elliptical galaxy is formed. There is ample evidence for smaller-scale interactions with the Milky Way, however—some of which are still happening today.

For example, evidence has recently emerged that the globular cluster Omega Centauri might be the core of a dwarf galaxy that has been subsumed by the Milky Way. This spectacular cluster, which is a fabulous sight through a small telescope, contains three generations of stars. This is highly unusual for a globular cluster, which normally contains a single generation of very old stars that were born together and will live and die together. To some astronomers studying the object, the three generations suggest that Omega Centauri may just be the remnant of a dwarf satellite galaxy that interacted with the

Milky Way during a period of intense bombardment, then settled to its current orbit in the galaxy's halo. As our ability to chemically fingerprint individual stars improves, we may be able to link halo stars and globular clusters to individual dwarf galaxies that merged with the Milky Way more than twelve billion years ago.

It is clear that the galaxy continues to grow today through interactions with smaller satellite galaxies and gas clouds. There are several distinct streams of stars and gas orbiting our galaxy at various wild angles outside the flat plane of the disc. The Sagittarius stream, for example, is a living relic of the ongoing interaction between the Milky Way and the Sagittarius dwarf galaxy, which lies further out from the globular clusters, beyond the Milky Way's halo. This band of stars stripped from the satellite galaxy is the product of a celestial tug of war billions of years in the making.

Although it lies very close to the Milky Way, the Sagittarius dwarf elliptical galaxy is hidden on the far side of the centre of our galaxy's core. Due to all the dark dust towards the centre of the Milky Way, it was all but obscured until its discovery in 1994. The path of the Sagittarius dwarf elliptical galaxy around the Milky Way would have taken it through the disc of our galaxy several times in the past. There are hints of the damage it did as it (and its hefty dark matter halo) interacted with our densely populated galactic disc. Computer simulations show that its repeated passage through the Milky Way's disc could even have caused the shock waves that we see today as the spiral arms.

There are several other galactic streams associated with various satellite galaxies and globular clusters, but one is very different as it is composed entirely of gas. It's called the Magellanic stream and is named after the small and large Magellanic clouds—two satellite galaxies that are readily visible with the naked eye from Southern Hemisphere skies.

You can spot the Magellanic clouds from latitudes south of 20 degrees above the equator, although you'll get the best view from south of the equator. They look like two spectacular clouds of stars om the main band of the Milky Way. To find them, first ghtest star in the entire sky, Sirius. You can check you've got t one because you will see the three stars of Orion's belt (which is surrounded by a rectangle of bright stars) pointing directly to it. Following the imaginary line between Sirius and Canopus—the second-brightest star in the sky—you will see a diffuse cloud of light about twenty times the size of the full moon between the constellations of Mensa and Dorado. This is the large Magellanic cloud. It looks like a small patch of the Milky Way that has become separated and floated off into the night sky. Further along that imaginary line, in the constellation of Tucana, you will find the small Magellanic cloud, which appears about half the size again. (There are several free smartphone apps that can help you locate objects in the sky—I use SkyView, which I highly recommend.)

Visible to the naked eye, the Magellanic clouds have been known for millennia. The first written records noting the clouds were made in 964 CE by the Persian astronomer Abd al-Rahman al-Sufi in his *Book of Fixed Stars*. They were later named after Ferdinand Magellan, a Portuguese sailor who set off with a fleet of ships on an expedition on behalf of the Spanish crown in 1519. The voyage ended early for Magellan, who got his comeuppance when attempting to impose colonial rule on the people of the Philippines—he was killed in a battle with a resistance force led by local chief Lapu-Lapu, who fatally wounded him with poison-tipped bamboo. By the end of the voyage, in 1522, only eighteen members of Magellan's expedition were left alive. These bedraggled survivors claimed credit as the first people to successfully circumnavigate the globe.

The large Magellanic cloud has only 10 per cent as much mass as the Milky Way. It has a strange morphology, with a pseudo-spiral disc structure, and it is sometimes generically described as an irregular galaxy, since its galactic bar is offset from the centre of the galaxy and its single spiral arm is lopsided. It contains ample gas and molecular clouds and is the site of some vigorous star formation. The small Magellanic cloud is a dwarf irregular galaxy that is long and thin in shape with many arcs, fans and filaments of gas pulled by the gravitational interaction with its neighbour.

The Magellanic clouds were probably formed at the same time as the Milky Way, but they have evolved their own stars—which means the gas in these galaxies has different chemical properties. For example, the large Magellanic cloud has only 50 per cent and the small Magellanic cloud has only 10 per cent of the heavy chemical abundance of the Milky Way. Hence, studying the birth and death of stars in the Magellanic clouds is a very good way to understand how environmental factors affect stellar evolution.

We have known for more than fifty years that the Magellanic clouds are linked by an invisible bridge of gas, which was discovered in 1963 by radio telescopes that detected emission from its cold hydrogen gas clouds. The bridge's origin may be in a collision between the small and large Magellanic clouds some 200 to 300 million years ago. It is part of an enormous looping arc of gas called the Magellanic stream, which trails the small and large Magellanic clouds in their circuitous orbit around the Milky Way. The Magellanic stream contains both cold and hot ionised gas and stretches over 200 degrees of the sky. That's more than ten times the size of Orion's rectangular body made up of the supergiant stars Betelgeuse, Rigel, Bellatrix and Saiph.

The Magellanic stream was formed by interactions between gas in the Milky Way and the gas of the Magellanic clouds, which was slowly stripped off as it pushed its way through our galaxy's halo. By staring at distant galaxies millions of light years beyond the Magellanic stream, astronomers have used an ultraviolet spectral analysis instrument on the Hubble Space Telescope to measure how much light is absorbed by various chemicals in the gas that makes up the stream. The results are clear: some filaments within the stream match the chemical make-up of the large Magellanic cloud, and other regions closely match that of the small Magellanic cloud. This demonstrates that the stream is composed of gas pulled by gravity from both satellite galaxies.

Perhaps the greatest scientific significance of the Magellanic stream is its future interaction with the Milky Way. The Magellanic stream carries more gas than is contained in one billion suns. If this gas is eventually added to the bulk of the Milky Way's disc, it could potentially sustain star formation in our galaxy for hundreds of millions of years. From its motion through space we can see that the Magellanic stream is currently falling towards our galaxy's disc at a rate of 10 million trillion trillion kilograms per year. What we're not so sure about is how much of its gas will actually reach the disc and how much will dissipate as it interacts with the hot gas of the Milky Way's halo. If the Magellanic stream disintegrates slowly as it interacts with our halo, or if it has a magnetic field that helps to knit it together, that might just be enough to enable it to survive the journey to the Milky Way's disc.

The Magellanic stream is not the only gas cloud interacting with the Milky Way: there is another class of objects called *high-velocity clouds*, which are bands of cold hydrogen, completely devoid of stars. They float high above the disc in our galaxy's halo and some are raining down on us at frightening speeds. The Smith cloud is one such high-velocity cloud and it is falling towards the galactic disc at 73 kilometres per second. Named after its discoverer, then Leiden University PhD student Gail Smith, it stretches 10,000 by 3000 light years in size (that's the size of a dwarf galaxy) and contains more gas than a million suns. As Smith's cloud falls through the interstellar gas of the Milky Way, its leading edge is chiselled into a bullet shape and its trailing material is swept back like a comet's tail. This thing is flying through space like an excited dog with its head leaning out the car window and its ears flapping in the wind.

Where does it comes from exactly?

We don't know for sure. The discovery of this fascinating object has led theoreticians to devise various origins for it. It could be gas left over from the formation of the Milky Way or stripped from the Magellanic clouds. Many have suggested that it is the gas component of the Sagittarius dwarf elliptical galaxy, separated from the stars as it was torn apart. But very recent measurements of the gas's chemical make-up show that its origin is probably closer to home, in the Milky Way.

My favourite explanation for the origin of the Smith cloud is that it results from a 'galactic fountain' originating in the Perseus spiral arm. The fountain, caused by a barrage of supernova explosions within a cluster of massive stars that all died in quick succession, drove a hot plume of gas way beyond the upper parts of the disc, which cooled and condensed into the Smith cloud as it fell back towards the disc. Other scenarios for the Smith cloud's origin will no doubt be investigated in coming years as astronomers find and piece together new evidence about its composition and trajectory.

What future for Smith's cloud?

Its current trajectory has been calculated accurately but its future is not entirely certain due to the complexity of the environment of the Milky Way's halo. One thing is clear: on its current course, it is due to hit the Milky Way's disc in twenty-seven million years' time, and when it does, it might just make some waves.

What can we hope for the future of the Milky Way? One specific question that astronomers are trying to answer is this: is enough gas falling down from the halo and feeding the galaxy's disc to sustain the current rate of star formation in the Milky Way? Or will the Milky Way run out of gas and suffer a starless future?

We observe many high- and intermediate-velocity gas clouds in the Milky Way's halo, and many of these are expected to settle down onto the Milky Way's disc. The total amount of fresh gas (hot and cold) estimated to join our galaxy each year is just over five solar masses—enough to sustain star formation at the current rate. The trouble is that a large fraction of the gas seems to be hot and ionised (that is, the atoms are separated into positively and negatively charged parts). That's bad, because gas in such a state cannot settle down to form stars. The fraction of cold gas approaching the disc from the galactic halo is too small to explain the rate of ongoing star formation that we observe, so we need to find some mechanism that allows the gas to cool down as it falls onto the disc—but we don't know how that could happen. Again, it is the job of theoretical astrophysicists to come up with accurate simulations to solve this problem and hopefully reassure us about the future of our galaxy.

Another source of clues about the history of our Milky Way is the supermassive black hole at its centre. The black hole and its bright surrounding material, called Sagittarius A*, are currently in a state of slumber. The black hole typically eats less than a 100-millionth of a solar mass per year, which is pretty slow going. But we know that this rate varies, since the accretion disc around Sagittarius A* (which is the first region to signal any changes in accretion rate) goes through frequent changes in brightness.

The X-ray emission from Sagittarius A* is slow and steady before bright flares occur several times a day. During these flaring events the X-rays become up to fifty times brighter than the ambient level. The short timescale suggests that the flares are coming from a region close to the black hole, since if they were further out it would take far longer than one or two hours for the material causing the flare to fall across the event horizon and disappear. Infra-red emission from Sagittarius A* also varies on timescales of hours and days. The proposed mechanism for this emission is electrons in the accretion disc being accelerated to speeds approaching a significant fraction of the speed of light. This might occur when a clump of gas drops into the black hole.

It seems that the variability of Sagittarius A* shows us in near-real time the conveyor belt of material falling past the point of no return and being digested to grow our galaxy's supermassive black hole. We can sometimes see this material approaching the black hole from outside. Since its discovery in 2011, astronomers have been tracking a massive gas cloud called G2 in its elliptical orbit around Sagittarius A*. As it has moved closer to the black hole it has begun to elongate and stretch, causing many researchers to think it will break up at its next closest approach.

There was a flurry of excitement in March 2014 as G2 neared the closest point in its 300-year orbit around the black hole, coming within 40 billion kilometres of the black hole's event horizon. During that tense week it sailed within 270 astronomical units (1 astronomical unit is the distance between Earth and the sun), or approximately the diameter of the solar system, of the black hole's edge.

Many thought this close passage might cause the cloud, which has an estimated mass of three Earths, to heat up, give off X-ray flares and finally rip apart, spilling all its gas over the event horizon and causing a flare of emission that we could watch in real time.

Astronomers watched and waited with telescopes poised as the cloud skirted the black hole's danger zone. In this game of cosmic 'dare', G2 was accelerated to *5400 kilometres per second* by the black hole's tremendous gravitational field. Researchers had predicted that there would be a large increase in the brightness of Sagittarius A* as a result of the event, with gas spilling onto the accretion disc and swirling down towards the black hole.

Much to everyone's surprise, G2 emerged from this stunt pretty much unscathed. Researchers monitored the X-ray, infra-red and radio emission before, during and after the event. The large increase in infra-red emission, predicted to rise as the gas cloud was compressed by gravity, did not eventuate. A steady increase in X-ray emission was also predicted but missing from observations, and an increase in the radio emission from Sagittarius A* during the event, which was predicted to occur as the bow wave in front of G2 passed Sagittarius A*, was not seen.

Flaring activity from the G2 passage had also been predicted and was monitored by various X-ray, infra-red and radio telescopes. Several flares were detected during the passage of G2, but the overall number of flares was pretty much what would be expected during a normal week. The number of medium-brightness X-ray flares (thought to be caused by the break-up of small fragments of rock or by the disruption of magnetic fields by material entering the accretion disc) actually decreased during the passage of G2. The only hint of increased activity associated with the event came in the form of two monster X-ray flares that were seen within a few months of G2's flyby, and the rate of very bright flares being higher than normal around the time of G2's close approach. However, these sorts of events happen from time to time and there is no direct evidence that they were connected to G2. Continued monitoring of the Milky Way's centre is now crucial to help us understand how frequently these extremely bright flares happen and whether the rate is reducing to background levels.

So what can we learn from these observations?

First and foremost, all the behaviours that were predicted for a gas cloud passing close to our black hole did not eventuate with G2's journey to the edge. Either all our physics knowledge is wrong, or G2 is not a gas cloud after all. With the benefit of hindsight, and armed with careful observations, several teams of astronomers created computer simulations of the event to try to explain the lack of fireworks as G2 visited the heart of our galaxy. The key to this process was to figure out how G2 could have held together so well in the face of such incredible gravitational forces.

Astronomers now agree that the most probable explanation for its great escape is that there is a star enshrouded inside the cloud of gas, and that this star was the strengthening bond that kept the cloud intact. This hypothesis is far from proven (we have still not seen the star), but it seems to be the best theory we have. Since the next close approach is not for another 300 years, that's probably the best we can do. There are many groups of astronomers still monitoring the galaxy's centre and waiting for the next opportunity to see our black hole snacking. Until that time, it's a waiting game to catch Sagittarius A★ in the act.

Despite the daily flares and the occasional close call with a star or a cloud of gas, our local supermassive black hole is a fairly quiet place. The slow and steady accretion rate at the galaxy's centre means that radio emission from Sagittarius A★ is only 1 per cent as luminous as the core of an active galaxy like Centaurus A. But mounting evidence shows it was not always this way.

Imagine you could somehow place a mirror far away in the depths of space, then shine a torch towards it. The mirror is so far away that the beam of torchlight travels 50 light years before bouncing off it and travelling another 50 light years back to Earth. In such a scenario, by looking in the mirror we would see the torch beam as it was emitted 100 years ago. This is a clever sort of time machine. Scientists have used this technique to study the past light output of Sagittarius A*. Instead of using a mirror, they have studied a reflection nebula close to the galactic centre, where X-rays are naturally reflecting off gas clouds that are floating around in space. The researchers have found light echoes from past bright flares reflecting off several adjacent molecular clouds in the Sagittarius region. In this way, we can see the delayed image of the flares that have occurred in the past, giving us an insight into the recent history of Sagittarius A*'s feeding activity.

The X-ray emission originating from our galaxy's central black hole and reflected off these clouds varies in its brightness on timescales between two and ten years. Interpreting the echoes is difficult due to the uncertainties in the distances between Earth, Sagittarius A* and the molecular clouds, but by refining current measurements of these distances, we may in future be able to gain further insights into past variability of the emission from the black hole, enabling us to piece together its response to feeding events.

Recent discoveries provide mouthwatering clues to the Milky Way's violent past. The most exciting and unexpected of these are the Fermi bubbles: two symmetrical, teardrop-shaped bubbles emanating from the galactic centre. Each bubble stretches 25,000 light years, covering an area of sky 50 degrees in diameter above and below the galactic plane. The term 'bubble' is a bit of a misnomer—they are not voids but more like balloons filled with electrically charged gas. The bubbles were first noticed in 2003 in microwave maps of the sky and were confirmed by the Fermi Gamma-ray

Space Telescope in 2009, which combined several years of data to make out the faint bubbles above the hubbub of gamma rays from our galaxy. Their origin is not known for sure, but what we do know from the presence of gamma rays is that they are powered by high-energy subatomic particles.

One process that could have created the Fermi bubbles is the accretion of large quantities of gas onto the disc of a supermassive black hole. When the black hole bites off more than it can chew, the accretion disc quickly becomes overheated. Some of the hot, electrically charged gas spills over and is violently ejected along the north–south magnetic field axis. This creates two gigantic streams of energetic gas flowing out above and below the disc of the Milky Way and spreading out as they go to form glowing regions like two lightbulbs shining out into deep space.

Images made with Earth-orbiting X-ray telescopes show regions of glowing-hot plasma at high galactic latitudes around the boundaries of the Fermi bubbles. The speed of the wind blowing the bubbles and their size give us a clue as to how long ago they started to form. The velocity of the outflow has been determined observationally by measuring the Doppler shift of gas entrained in the wind. It's blowing at a squally 900 kilometres per second (that's more than 2000 times the speed of sound). Using this speed and other evidence, most estimates of the age of the bubbles come in at between three million and fifteen million years. That might sound like they formed a long time ago, but in the cosmic scheme of things it is incredibly recent.

There is some debate about this interpretation, with alternative explanations for the bubbles being put forward. Some astronomers think that rather than being ejected by the disc of Sagittarius A*, the Fermi bubbles may have originated from a galactic wind generated by large numbers of hot massive stars that formed at the centre of the galaxy within the last ten million years. This 'Milky Way starburst' scenario would have caused large numbers of supernovae to go off in quick succession (that is, within a few million years of each other). The 'Milky Way starburst' origin is certainly plausible—after all, there are currently large associations of massive stars towards the centre of the galaxy and plenty of supernova remnants still visible.

Whatever triggered the formation of the Fermi bubbles, it was likely some form of impact, either with a gas cloud or a satellite galaxy. The material from the impacting object funnelled towards the centre of the galaxy and created a sudden burst of star formation or a major black hole accretion event, with 10,000 suns' worth of gas falling helplessly into Sagittarius A*'s lair. Personally, I prefer the black hole disc explanation for the origin of the bubbles. The idea of a black hole consuming 10,000 suns of gas and burping plasma bubbles into deep space seems rather exciting.

However they were formed, the discovery of the Fermi bubbles presents to us an intriguing question: is this evidence that the Milky Way was recently an active (or a starburst) galaxy?

Mounting evidence may be pointing us to a very active recent history for our galactic nucleus. Computer simulations of a fast jet of electrons and neutrinos from a supermassive black hole's disc have successfully re-created these fascinating features. Recent observations using the Herschel satellite have further compounded this interpretation, uncovering a previously unknown hollow ring of molecular gas surrounding Sagittarius A* that closely resembles rings of gas that are found in active galaxies, hinting that our galaxy may have been active in the recent past. Research will continue on this topic for many years to come as our understanding of the Milky Way's past is slowly unravelled—so watch this space.

It makes sense to wonder whether this activity could start again. After all, we know that material continues to orbit the black hole in close proximity and will continue to join Sagittarius A* at periodic intervals. But are there larger volumes of material waiting to fall in, as the 10,000-solar-mass cloud apparently did in the past ten million years to create the Fermi bubbles?

Given the dynamic environment of our galaxy and its frequent and ongoing interactions with satellite clouds and dwarf galaxies, it seems likely that the answer is a resounding yes. With the amount of satellite galaxies and gas clouds raining down on the Milky Way's disc, it can't be long before our galactic engine revs up once again.

Activity in a galaxy's nucleus causes a number of immediate effects on the environment, the most obvious being the generation of enormous jets emanating from the galactic centre. Jets are narrow beams of heat made up of fast-moving particles called *cosmic rays* and ultra-violet, X-ray and potentially gamma-ray radiation. Although we are protected from normal levels of these types of energy by our atmosphere and magnetic field, higher levels could be harmful to human health and pose risks to all life on Earth. So if Sagittarius A* did experience another major accretion event onto the supermassive black hole like the one that generated the Fermi bubbles, could it affect us here on Earth?

In 2017, astrobiologists from the University of Rome Tor Vergata published a paper in the journal *Nature* reporting on a study of the implications for habitability of Earth of heightened activity at the galactic centre. They looked at the likely effect of a major black hole accretion event and its associated energy outburst on our atmosphere and through our direct exposure to radiation. Their results paint an uncertain picture for the long-term future of life on Earth.

If it were exposed to extreme levels of ultraviolet radiation, our atmosphere could potentially be stripped—literally evaporated—from the surface of Earth. Whether this happens will be down to the exact dose of radiation that Earth receives and depends on a number of factors: the brightness of Sagittarius A*, our distance to the galactic centre, whether the radiation is confined to a jet or shining in all directions, whether any potential jet is pointing towards us, and how much shielding we receive from cold molecular gas surrounding the galactic centre. With so many unknowns, the uncertainties in these estimates are quite large!

Some of the factors are very much on our side. First, we live in the galactic plane, where protection from the doughnut-shaped cloak of molecular gas encircling the Milky Way's centre is at its highest. Second, we are located towards the outer regions of the galaxy's disc, 25,000 light years away from the supermassive black hole—this distance gives us significant protection as it dilutes the ionising radiation emitted from the centre of the Milky Way. There is a school of thought that we live in the galaxy's 'habitable zone', which is shielded from such outbursts of activity in the Milky Way's central engine.

Planets towards the centre of the galaxy may not be so lucky, running the risk of getting sprayed with dangerous radiation from events such as supernovae, gamma-ray bursts and black hole burps. Within about 3000 light years of the Milky Way's centre (that is, within the galactic bulge), an Earth-like planet would stand to lose a very significant fraction of its atmosphere as it was evaporated by the hot blast from the black hole accretion event. Planets within 1500 light years of the Milky Way's centre would see their atmosphere disappear completely if such an event were to occur.

Even though our distance from the centre protects our atmosphere from evaporation, ionising radiation is still capable of damaging the ozone layer. With a partially damaged atmosphere, the radiation exposure from a sudden and powerful flare from Sagittarius A* could potentially wipe out all multicellular life on Earth, including humans. Even a slow and steady accretion event lasting hundreds of thousands of years is likely to give us a hefty (but non-lethal) radiation dose. Chronic exposure to ionising radiation is known to cause cancer, cataracts and potentially dangerous genetic changes. Some prokaryotes (single-celled organisms) have extraordinary resistance to extreme radiation and would likely survive the onslaught, but complex life forms could be seriously impacted or forced to adapt.

So will this doomsday scenario play out in our galaxy, leading to the sterilisation of our planet, the cradle of all known life in the universe? Will we all be growing three heads anytime soon?

A number of factors seem to point to the answer being a reassuring no. After all, our galaxy has been through such an accretion event before, in the past ten million years, and this event (which created X-ray jets and the Fermi bubbles) demonstrably failed to wipe out our predecessors. Plus we live nestled in the suburban region of our galaxy's disc, which is quieter than the crowded and turbulent inner regions.

One thing is for certain: in a galaxy that created us but is sometimes out to kill us, all that is separating humans from extinction is a magnetic field and a thin layer of air. We should at least be grateful for that.

7

MEET THE NEIGHBOURS

Gravity is a puny master. All things being equal, it is the weakest of the four fundamental forces of nature, the others being electromagnetism, the strong nuclear force and the weak nuclear force.

Electromagnetism is the reason that objects have their shape. It is the bond that clasps your fridge door shut and it binds together the electrons and protons in every atom in your body. It is the attraction between positive and negative and it draws a compass needle north and south. The electromagnetic force works on all scales, from the smallest subatomic particle to the biggest galaxy. Its influence flows freely, like gravity, with the strength of the force fading exponentially as you get further from the source.

The two nuclear forces, on the other hand, are only felt by objects that are in unimaginably close proximity.

The strong nuclear force is what binds together the smallest fundamental particles, called quarks, to make protons and neutrons. It also bonds the protons and neutrons to form atomic nuclei. These are the fulcrum of the atoms, marking the collections of matter in the latticelike structure of all materials on Earth. Without it, there would be no protons, neutrons or even atoms. To this force we owe our entire existence.

The weak nuclear force is another short-ranged interaction, but it occurs on scales less than the diameter of a proton (that's 1000th of a trillionth of a metre, in case you're wondering). The weak force has the ability to transform neutrons into other types of fundamental particles, enabling radioactive decay by the process of fission. This is the basis of nuclear power, which (love it or loathe it) generates 14 per cent of the world's electricity. The weak force can also cause protons to change into neutrons, which, in tandem with the strong force, can enable nuclear fusion. This is the process in the cores of stars that lights up our universe, creates all the chemicals required to sustain life, and provides the warmth and light that enable life to exist on Earth.

Gravity, the fourth and final force, is what glues us all to the ground and keeps Earth orbiting the sun and our solar system cruising around the Milky Way. It takes a lot of matter to generate a small gravitational force. That's why, on the scale of a single atom, gravity is incredibly weak. In a hydrogen atom, the electromagnetic force required to pull the negatively charged electron away from the positively charged proton is 0.00000008 Newtons. To break the gravitational bond between a proton and an electron in a hydrogen atom requires a force somewhere in the region of 0.0000000000 00000000000000000000000000000000000036 Newtons. To help calibrate those figures, the gravitational force between my body (with a mass of just under 60 kilograms) and the surface of Earth is around 580 Newtons. The strength of gravity on the scale of an atom is also completely insignificant compared with the strong and weak nuclear forces.

But despite the relative weakness of gravity, it is the dominant force at large scales in our universe, assembling all the large-scale structures we see. How can that be?

The answer is that although the strong and weak nuclear forces are dominant on small scales, their influences don't stretch very far.

So on the scales that we experience in everyday life, the nuclear interactions don't get a look-in. The electromagnetic force, although it is far stronger and has a long reach like gravity, has attractive and repulsive elements. If you add up all the like-charged particles (positive-positive and negative-negative charges) in the sun, they repel one another through the electromagnetic force. The oppositely charged particles, on the other hand, attract one another to an equal degree as that of the repulsive forces. What is the net electromagnetic force on the sun? Zero.

Gravity is a force that is always attractive. There is no 'anti-gravity' force. If you add up all the particles in the sun, they all attract one another through gravitational force with no regard for the electrical charge of the particles. In fact, if I managed to stand on a special insulating platform close to the sun's gas 'surface', a set of bathroom scales would show that I weighed 1.6 tonnes. That is not because I have added any bulk to my body—it's just the consequence of the much larger attraction of my body's mass to the mass of the sun, compared with the mass of Earth. The net gravitational force within the sun, which is fully on 'attract' mode, wins out over the ostensibly 'stronger' electromagnetic forces, which have attractive and repulsive elements.

Gravity has an influence that far exceeds its diminutive power. It's the gravitational force that renders the sun a sphere and keeps us in our year-long orbit. It also keeps our galaxy locked into a dangerous dance that could eventually spell the end of life on Earth.

Matter in the universe—even on its biggest scales—is not distributed evenly but is highly clustered. Dark matter links together chains of galaxies like a gigantic jigsaw puzzle, bringing with it all the galaxies that we observe. Called the cosmic web, this complex and delicate filamentary structure underlies everything we see, as well as the dark matter we can't see. The supernatural spider that has woven this web, arranging billions upon billions of galaxies into giant clusters and leaving enormous voids in space where almost nothing exists, is the force of gravity.

The scale of the cosmic web is truly humbling. Gigantic clusters of galaxies like the Virgo cluster, in which the Milky Way resides, have diameters of between 7 and 30 million light years, and are linked by narrow filaments of galaxies more than 30 million light years long. Voids, which are almost completely devoid of galaxies, are of a similar size to galaxy clusters and link these structures. Thinking about those sizes almost makes my nose bleed. Light travels at a speed of 300,000 kilometres per second—a speed that would see it dash around Earth's equator more than seven times per second. For a ray of light to zoom from one galaxy cluster to the next would take the same period of time that occurred between the disappearance of Tyrannosaurus rex and the appearance of the Spice Girls. That is how dizzyingly large our universe really is.

Astronomers are mapping out the large-scale structure of the universe by carrying out huge surveys of thousands of galaxies across the night sky. Recent surveys include the 6-degree-Field Survey, carried out by the Australian Astronomical Observatory with the 1.2-metre-diameter UK Schmidt Telescope at Siding Spring in Australia, and the Sloan Digital Sky Survey, undertaken with a 2.5-metre-diameter telescope at Apache Point in New Mexico in the USA. These great projects, which take several years of work by large international teams of scientists to achieve, have brought the structure of the cosmic web more clearly into focus. The resulting images show an intricate web of galaxies stretching across hundreds of millions of light years. Computer simulations of the large-scale structure of the universe match incredibly well with the observations, telling us that our understanding of the processes leading to their formation is pretty close to the truth.

The origin of this large-scale structure is the tiny quantum fluctuations in density that arose in the first fleeting seconds after the Big Bang. Quantum mechanics explains our universe in a strange way, with tiny particles never sitting still but, rather, continually popping in and out of existence in very similar locations, giving the impression from our point of view of being static. Strange it might be, but it definitely works, and quantum mechanics is able to explain and predict the behaviour of matter on small scales very well. Early in the universe, then, the tiny fluctuations in density caused by random jumps by these tiny particles were enhanced by the rapid expansion of the universe; since then, the condensation of dark matter by gravitation has started to do the rest. Large-scale gravitational collapse of dark matter seeded the formation of galaxy clusters and a complex network of smaller structures. With the oppositional forces that regular matter is subject to (for example, pressure), it took somewhat longer for this regular matter to collapse into individual galaxies. But as it did, the universe as we know it began to take shape.

Although the cosmic web is very evident in the arrangement of galaxies, if you expand out to even larger scales (greater than about 300 million light years) the detail is diluted and the matter in the universe seems pretty smooth. You can imagine this like a wooden tabletop. On the face of it, you can run your hand up and down the surface of a table and it feels pretty smooth. But if you shrink yourself down to the size of an ant, you can run up and down the tabletop and start to notice the roughness of the wood and the small scratches and knots. Compress yourself even further, down to a microscopic scale, and you'll see the fundamental structure of the wood—chains of carbon-based molecules arranged into long, thin tubes with large air gaps between them. This is the beautiful structure that gives wood its grain. The wooden table is no longer smooth.

The universe is the same. On large scales it appears homogeneous—it is smooth and looks pretty much the same in every direction. But when you get into the detail, it is very intricate indeed. The voids in the cosmic web, down to clusters and groups of galaxies, the Milky Way, stars, planets, cities, people, atoms, electrons, quarks … these are all levels of detail in our cosmos that are being experienced on increasingly smaller scales. And the difference between the largest (a void in the cosmic web) and smallest (a quark) scales in our universe is around fifty orders of magnitude.

When I think about this mind-boggling range of physical scales in the universe, it makes me realise that our experience of our world—and indeed the universe—is highly dependent on relative size. If life existed (and why couldn't it?) on the scale of galaxy clusters rather than being huddled as we are on the surface of one planet, those creatures would barely know that planets exist. Perhaps there are unimaginably large living things in the universe that are harvesting the chemicals, light, magnetism, gravity or nuclear power from their surroundings to fuel their existence. Maybe they are composed of dark matter, if it exists, or perhaps they have no physical form at all. After all, our bodies only look as they do because we have evolved over millions of years to live on the surface of planet Earth. We may have no means to see these creatures or interact with them. To them, we would be like the microscopic dust mites that we ignore in our everyday lives. Perhaps we're looking for extraterrestrial life in the wrong places?

Our home within this grand design is in a group of galaxies called the local group. It's a relatively small accumulation of fifty or so galaxies stretching 10 million light years across. On a broader scale, our local group is one of hundreds of galaxy groups and several galaxy clusters that make up the Virgo supercluster. Our local supercluster spreads across more than 100 million light years, contains almost 50,000 galaxies and, as we will see later, is just a tiny part of even larger structures.

The local group is dominated by a trifecta of large spiral galaxies: the Milky Way, the Andromeda galaxy and the Triangulum galaxy. The largest of these is Andromeda, which you can easily see with the naked eye from a reasonably dark place in the Northern Hemisphere. To find it, first locate the bright 'square of Pegasus'. From the top left star in the square, follow the line of stars that runs diagonally upwards from it until you get to the second star, which is called Mirach. Now jump one star to the right, then jump the same distance to the right again. The Andromeda galaxy is visible as a fuzzy smudge of light.

As a teenager, this was one of my favourite objects to look at in the night sky, not least because it was extraordinary to think that I was looking at something—with the naked eye—that was 2.5 million light years away and made up of a trillion stars. This tiny, puny little fuzzy ball of light was the combined output of light of 1000 million suns! Not only that, but its distance gave me a direct view of the Andromeda galaxy from an era when Homo habilis walked Earth, and before our human ancestors learned to use fire. I always imagined someone peering back at me in 2.5 million years' time, wondering what creatures now exist in our galaxy.

Technology has moved on and nowadays you can locate the Andromeda galaxy using a free smartphone app such as SkyView. If you take a look through a pair of binoculars, you'll see that the light actually has a shape and is made up of a bright core and a fainter, more diffuse elongated disc. This is the Andromeda galaxy, the closest spiral galaxy to the Milky Way. It contains approximately four times as many stars as the Milky Way and a gigantic gas halo 2 million light years in diameter. If the Milky Way's disc has a similar-sized halo relative to its number of stars, we might expect that the two haloes are touching or possibly even overlapping in space. Andromeda may have more stars than the Milky Way, but the two galaxies have a similar mass, with ours containing far more dark matter. Nobody knows why. Around the Andromeda galaxy, like the Milky Way, is a swarm of dwarf galaxies and other irregular satellite galaxies that are bound to Andromeda by gravity.

A much smaller proposition, the Triangulum galaxy is a barless spiral galaxy containing around forty billion stars and is less than half the size of the Milky Way. It forms a triangle with the 'big two' galaxies, lying around 3 million light years from the Milky Way and 750,000 light years from Andromeda. Astronomers have calculated a rough speed and direction of motion for the Triangulum galaxy by looking at the radial velocity of a bright cloud of water molecules residing within the galaxy. Triangulum seems to be moving towards the Andromeda galaxy, which could mean that it is captured in an orbit around the much larger spiral galaxy. More research is needed on this particular point, since the measurements are uncertain at present and tracking the water masers over a series of several years will help to pin down the accuracy of the orbit.

The Triangulum galaxy is visible in northern skies quite close to (within a few degrees of) the Andromeda galaxy. It lies in the constellation of Triangulum, which is a lazy group of just three stars that was named by Ptolemy in the second century. Unlike the other great constellations of the north depicting lions and hunters and mythological water bearers, it is simply a triangle of three rather average stars. Ptolemy was obviously having a slow day at the office when he came up with this particular constellation. It does contain this challenging galaxy, however, which is the most distant object in the universe visible with the naked eye. You'll have to travel to a really dark place to see it unaided, and with light pollution affecting our eyes' ability to distinguish such objects, it's pretty difficult. I have never managed it. Through a small telescope borrowed from my pals at the local astronomical society, though, I once saw the Triangulum galaxy's elongated form from my childhood back garden in England.

The remaining fifty-plus members of the local group are mostly dwarf or irregular galaxies. Dwarf galaxies are generally very faint, and many have only been discovered recently, as surveys with new and sophisticated telescopes carry out very careful checks on the relative numbers and motions of stars in different parts of the sky. The smallest and faintest found so far is less than 1 per cent the size of the Milky Way and only as bright as 40,000 suns. I'm sure that many more of these tiny galaxies will be discovered over time.

A few things are known about the origin of these minor galaxies. Observations of dwarf spheroidal galaxies, which are relatively faint and gas-poor, show that they are more common towards the centre of the local group (close to the Milky Way and the Andromeda galaxy) than in the suburbs. Irregular galaxies, on the other hand, in which gas is more plentiful, are more likely to be found towards the edges of the group. This observation supports the idea that these minor galaxies may have evolved in their shapes, sizes and gas content as they interacted with the large spiral galaxies at the centre of the local group. Not only can these interactions trigger the evolution of minor galaxies, but they are also capable of creating them, ripping off clumps of stars and gas and hurling them into space.

The Andromeda galaxy is accompanied by more than thirty satellite galaxies, including M110, an elongated dwarf spheroidal galaxy that is clearly visible as a thin smudge of light in telescopic images of Andromeda. It is orbitally linked to M32, a compact elliptical galaxy that contains a central supermassive black hole with a mass in excess of one million suns. This black hole, called M32* (pronounced 'em thirty-two star'), emits X-rays as it gradually consumes stars from the central region of M32. The galaxy itself is dominated by older, cooler yellow and red stars and contains very little star-forming material. The ejection of the galaxy's gas may have been triggered by a past merger with another galaxy, either by the galactic winds associated with black hole accretion events or by mass star formation events in the distant past.

There are three dwarf elliptical galaxies in Andromeda's clutches, as well as a handful of dwarf spheroidal galaxies. These, along with M32, M110 and possibly other long-since-consumed satellites, are having a significant gravitational influence on the Andromeda galaxy. There are many tidal streams and clumps of material scattered around Andromeda. The largest of these is called the giant stellar stream and is the remnants of debris left over from a galaxy that came within a whisker of the Andromeda galaxy's centre around a billion years ago. It left a dent in the disc of Andromeda and a stream of debris in an orbit stretching 300,000 light years in diameter. It is also thought to have scattered stars and clusters around the inner region of the Andromeda galaxy's halo. Interactions such as this have led to the disc of the Andromeda galaxy being more warped than the Milky Way's, and showing lumps and bumps in many different places. This points to a frequent and significant bombardment by minor galaxies in Andromeda's past.

The Milky Way is accompanied by around thirty satellite galaxies, some of which are actively interacting with our galaxy as they pass close by in their orbits. The brighter examples, such as the Magellanic clouds, have been known for a long time since they are particularly obvious in the night sky, as I found out recently on a trip to the Siding Spring Observatory when I was filming a television show. As I emerged bleary-eyed from my room at 3 a.m., the darkness and sheer expansiveness of the sky hit me like a hammer. My eyes took a few moments to adjust, but as I craned my neck and looked directly above me, the Milky Way was laid out clear and glittering like a million distant flashbulbs from a red carpet full of paparazzi. The Magellanic clouds shouted their sparkling existence from the rooftops. This, folks, is real astronomy. My dumbfounded reaction was to stand with my mouth agape and repeat 'Wow' and 'Oh my God' over and over again. Sadly, it's rare for city-dwelling astronomers to see a truly dark sky.

The majority of satellite galaxies in our local group, including the Sagittarius dwarf galaxy and its associated galactic streams, are far too faint to be seen without specialist equipment. Recent discoveries include dwarf galaxies such as the tiny Bootes III, discovered in 2009 and thought to be the final remnants of a shredded dwarf galaxy associated with the faint Styx galactic stream. There is also a contender for the best official name in space, the Monoceros Ring, a structure of stars that is somewhat controversially thought to be a galactic stream around the Milky Way arising from the orbit of the Canis Major dwarf galaxy. Others believe it is actually a warp in the Milky Way's disc. Whatever its origin, the Monoceros Ring seems to be connected to the interaction of the Milky Way with another massive body.

The whole of the local group is bound by gravity, so each galaxy is influenced by the others. The Milky Way, as one of the more massive objects, has a significant gravitational influence on its satellite galaxies but is also at the mercy of dozens of smaller gravitational tugs from its neighbours, including the massive Andromeda and Triangulum galaxies. As a result, the Milky Way is locked in a tumbling orbit around the group that can be crudely measured by looking at the Doppler shift of stars within our neighbouring galaxies.

We still don't have a detailed three-dimensional picture of the motions of galaxies within the local group. In fact, the satellite galaxies are so faint and hidden by the dark material threading our galaxy's disc that almost half of them have only been discovered in the past decade. This means that our understanding of their properties and motions is an ongoing effort. Even for the well-known members of the local group, such as the Magellanic galaxies, obtaining an accurate measurement of the motion in the two directions towards/away from us (radial) and sideways (transverse) is incredibly difficult due to the tiny year-on-year motions involved.

Fortunately, new techniques and space telescopes like the Hubble Space Telescope, its successor the James Webb Space Telescope, and Gaia (which as it orbits Earth is systematically measuring the positions and sky motions of millions of stars) will bring more certainty to our understanding of the local group over the coming years.

We've known for some time that the Andromeda galaxy is generally moving towards the Milky Way, but information about its precise trajectory was missing from the picture. That was until 2012, when a team of astronomers used the Hubble Space Telescope to track the transverse motion of stars in the Andromeda galaxy relative to very distant background galaxies. The study revealed that their sideways motion is very small—in fact, the stars in Andromeda are moving almost directly towards us at an astonishing rate. The Milky Way and the Andromeda galaxy are flying headlong towards one another at around 400,000 kilometres per hour—that's 100 times faster than a bullet and ten times faster than a rocket launch. At that speed, we can expect a head-on collision between the two biggest galaxies in the local group in precisely 3.87 billion years' time. The consequences for Earth during this tumultuous event could be dire.

8

WHEN GALAXIES COLLIDE

When you're sitting around a campfire looking up at the stars, what do you see?

A background of inky blackness crossed by a silver spray of stars. Bright leaders dazzling and dancing in the evening air and maybe a crescent moon low in the sky to light the way. It is something that almost every human being has experienced and felt a sense of awe about. People have lived and died by stories of the sky, and science was born from attempts to make sense of its ways.

Our night sky is truly something that unites generations. It looked much the same when the civilisations of antiquity were flourishing, and back through the first agricultural societies in the Neolithic era. Even in the early Stone Ages several tens of thousands years ago, our world-wise ancestors would have shared a clear view of a blackened sky with the same Milky Way, the same moon and much the same star patterns that we see today. Imagine sitting down with them and (despite not sharing a language) pointing out the familiar patterns, displaced slightly by the glacial shifting of the stars within our nearby spiral arms, and sharing in a common thread that is woven through our collective memories.

Now imagine that our descendants could come back and visit us with their inexpensive time machines. What stories would we and 'future us' be sharing as we sat together on a hillside taking in our star-scattered view into the heart of the Milky Way? Would our visitors react with familiar awe or with bewildered wonderment at our dark, cavernous skies?

Given what we have learned about our place in the universe and our home galaxy and its adaptation to change, we can expect some pretty big changes in our cosmic outlook over the next few billion years. The gravitational shifts in our local galaxy group are bringing together the Milky Way and the Andromeda galaxy like two colossal tectonic plates on a head-on collision course. As we have seen in so many of our cosmic studies, galaxy collisions can bring about complete transformations of spiral galaxies—they can be warped, mashed and mangled as their discs tangle and compress, triggering mass star formation and lighting up the sky in a festival of light. As the cores of the two galaxies grow closer, the two supermassive black holes become entwined and slowly squeezed as material is scooped up and flung into space as super-luminous jets. Finally, the black holes merge with a 'chirp' and a pulse of gravitational waves. This is the journey our Milky Way will tread, as it would be observed at a great distance.

But what will it be like when it's our own galaxy caught up in a merger? How will it look from our planet? And how will these changes affect the prospects of life on Earth and other planets?

The collision is a creeping threat. Our assailant Andromeda is already in our sights, being faintly visible to the naked eye. Since it is so faint, its apparent size in the sky might surprise you. Until recently, estimates of the size of the Andromeda galaxy's disc were in the region of 70,000 to 120,000 light years across. The large range quoted is not because we're uncertain about measuring the size, but because the diameter of the galaxy depends on how you measure it. If you look in optical light you generally see the bright inner region of a galaxy's disc, whereas with a radio telescope you can pick up a lot more faint gas around the outskirts. A 2005 study with the Keck telescope showed that stars in the faint outer portion of the Andromeda galaxy's disc stretch at least 220,000 light years in diameter, which means the maximum angular size of the disc is close to 3 degrees (that's six full moons long). Sadly for us, Andromeda is quite faint, so not only is it hard to see without a telescope, but with the human eye we can see only the small bright portion of the galaxy made up of the stars in the bulge and the central region of the disc. Imagine looking up into the sky every night and seeing this amazing big spiral shape hanging among the stars. If only our eyes were a little more sensitive and attuned to this faint portion of the disc, it would look absolutely incredible!

As the Andromeda galaxy rushes towards us at 400,000 kilometres per hour, its apparent size will only increase with time. This change is not perceptible within a human lifespan, even though our two galaxies will be about 300 billion kilometres closer to one another at the end of your life than at the beginning. We would need to fast-forward more than a billion years to notice any major changes. By then, our solar system will have travelled about four and a half orbits around the Milky Way and the nearby stars will have moved and shifted, dissolving our familiar constellation shapes and replacing them with new ones. In 1.9 billion years' time, half the time taken to reach collision, the Andromeda galaxy will appear to have doubled in size and quadrupled in area, taking up a patch of the sky more than twelve by four times the diameter of the full moon.

The Andromeda galaxy will not only get bigger, it will also get much brighter. According to the inverse square law, as an object gets two times closer it gets four times brighter. So in two billion years the Andromeda galaxy will not only appear as large as the small Magellanic cloud does now, but it will be significantly brighter. In fact, it will be as bright as Regulus, the most brilliant star in the constellation of Leo.

Assuming that the scientific enlightenment of any species is never complete, anyone in the Milky Way who is witnessing such a spectacle will probably have important cultural and superstitious attachments to the Andromeda galaxy, big and bright as it is. It will probably be an object of particular significance in the night sky for civilisations throughout the Milky Way (or at least any part of the disc with an uninterrupted view). It may even be someone's god, or a vital character in a moral tale.

In about three billion years' time, the Andromeda galaxy will begin to dominate our night sky. By then, it will be only 570,000 light years from Earth and cover an area of sky twice the size of the palm of your hand. Hold both your palms out at arms length and take a look at how enormous it is! The spiral-shaped disc of Andromeda will be more than forty times the width of the full moon. Due to its ever-increasing proximity, its brightness will exceed that of Alpha Centauri, the third-brightest star in our sky. Civilisations right across the Milky Way, assuming they exist, will see Andromeda almost as another moon, a large and dominant presence in the sky. Its brightness and proximity will also be a boon to the astronomers of the time. Having another nearby galaxy to study in detail will provide fantastic opportunities to understand how the differing environments of galaxies relate to their properties and behaviours.

In three billion years' time, the shape and structure of the Milky Way and Andromeda galaxy will still be largely undisturbed. But sometime in the future, the gravitational forces between the two galaxies will pass a point of no return. Not only will they accelerate towards their final merger, they will also begin to bend and buckle under the strain. Detailed computer simulations of the interaction show this messy process starting in around 3.8 billion years' time.

To anyone watching the night sky, our Milky Way disc will look fairly normal at first, a glittering road paved with stars, but the sky will be blessed with a second band of light—the enormous disc of the Andromeda galaxy—angling out of the long strip of the Milky Way. This will be an astounding time for stargazers, with two galaxies putting on a show across our sky each and every night. As huge gravitational strain begins to stretch and warp both discs, they will begin to bend and the spiral arms to unfurl, making the Milky Way as an astronomical phenomenon appear curved and more clumpy and uneven than it does today.

The situation will evolve as the Milky Way plunges into the disc of the Andromeda galaxy, beginning around 3.87 billion years from now. As the discs of the two galaxies pass through one another like ghosts, the shock waves from the collision will dismantle the beautiful spiral arm structure of both, spreading two fans of stars far into intergalactic space. This moment will mark the end of the spiral structure of the two galaxies and the beginning of their long journey of amalgamation and evolution. Whereas now the Milky Way appears as a narrow, straight band of light across our sky, in this tumultuous era the night sky will be packed with giant arcs—partially unfurled spiral arms. Like interacting tentacles, they will embrace planet-dwelling astronomers like a giant cosmic octopus.

The collision will trigger widespread star formation in both galaxies, producing clusters of young, bright stars that will begin to dominate our night sky. This first pass will leave the Milky Way much brighter in places and sprinkled with large numbers of extremely bright blue stars, each burning 100 times hotter and more powerfully than our sun. The general light output of our Milky Way will become much greater, and the stars in our night sky will be brighter and more plentiful.

Many new stars will form in clusters, and if we're lucky we might even end up with a spectacular close-up view. A nearby cluster of young stars would appear like a souped-up version of the Beehive cluster in Cancer, the Trapezium cluster in the Orion nebula, or the Pleiades (aka the Seven Sisters), which are all dazzling through a telescope. A sky filled with jewels like these would be a mesmerising sight.

Such beauty will come with peril for the observer, however. These clusters will contain large numbers of very massive stars that will burn through their hydrogen fuel in a few million years.

When the fuel is spent, the stars will go supernova—and close to an exploding star is a place you definitely don't want to be. If a supernova went off within about 50 light years from Earth, we would be bathed in lethal cosmic radiation that could disrupt or even destroy life. Gamma rays would trigger chemical reactions that would deplete our ozone layer, leaving us and our food sources immediately and critically vulnerable to damaging solar radiation. Life on Earth could be wiped out within a few years. Such are the perils of living in a starburst galaxy. Perhaps we'll take the telescopic view of star clusters after all.

If Earth were a little further from one of these clusters, when the massive stars went supernovae the gamma rays would not strip our atmosphere, but they would still irradiate our planet. This could cause changes to the genetic code of living creatures on Earth, including humans. Gamma rays are a well-known mutagen, altering the genetic code of living material. It sounds scary but isn't always harmful: deliberate exposure of seeds to low levels of gamma radiation is a common agricultural technique used to generate new species of plants that are resistant to frost, disease, salt or drought conditions. In the future starburst phase of our galaxy, the right levels of exposure to gamma rays could trigger a rapid series of genetic mutations that change the face of our world. Life on Earth would rapidly evolve, forever altering the balance of our ecosystems and leading to adaptations that we can't even imagine today. In a gamma-bathed galaxy, planets across Milkdromeda with primitive life would burst into a phase of rapid evolution like the Cambrian explosion that happened on Earth just over 540 million years ago, when thousands of species developed complex apparatus such as eyes.

What fate our solar system?

In the first pass of the two galaxies 3.8 billion years hence, many millions of stars will be flung from their cradles and left surging through the aether alone, like warp-speed-travelling spaceships. Depending on the position of the sun at the time, we will either be hurled outwards into a new, bigger orbit around the Milky Way, or we will keep going forever into outer space. If this is the lonely fate of the solar system, our descendants will begin to travel out from the Milky Way, seeing it recede and fade as time goes by until all the sky becomes uniformly dark. At this point, the only objects visible at night will be our fellow planets, the moon, and one or two other orphan stars that happened to be hurled out of the Milky Way in a similar direction. Eventually, any nearby stars will spread out and the sky will be filled with a profound darkness.

Imagine living on this future Earth, or a similar travelling planetary system flung far from the Milky Way, where all you can see at night is darkness, punctuated by one or two wandering planets. There would be no stories of the sky passed down through the generations, and no constellations to admire. Navigation at night would rely on reference to complex lunar or planetary tables rather than a quick glance at the northern or southern polar constellations. Would the concept of astronomy beyond our solar system even exist? If telescopes have been invented to look at distant objects on Earth, would anyone think to train them on the virtually featureless night sky and find the billions of unseen distant galaxies?

The chances of us leaving the galaxy entirely and meeting this 'dark sky' fate are thankfully very slim, but such an intriguing outcome will depend on the position and timing of the collision relative to our position in the Milky Way's disc, which at this stage is anyone's guess.

After the first interaction between the Milky Way and the Andromeda galaxy, the distorted remains of the two galaxies will slowly spin apart for 800 million years, causing Andromeda to fade slightly in the night sky but draw with it a band of stars and gas that will extend between the two galaxies. Next, as gravitational forces take over once more, the galaxies will be drawn back together for a second collision, causing their discs to pass through one another again in about five billion years' time. Each successive pass through the galactic discs puts the brakes on the relative motions of the galaxies, like a bouncing ball losing energy every time it hits the ground. This gravitational drag brings our two galaxies towards an eventual merger, which calculations show will happen on the third collision in approximately 5.86 billion years from now. During this phase of the merger, our galaxies intertwine, never to be separated again. Milkdromeda is born.

How will the other large spiral galaxy in the local group—the Triangulum galaxy—fare in this cosmic stoush?

The answer is that we can't yet be sure. Due to uncertainties in the current trajectory of Triangulum with respect to the Milky Way and the Andromeda galaxy, we can't be certain of the role of Triangulum in the Milkdromeda collision. The best astronomers can do is carry out a series of computer models and determine the outcomes of a variety of possible scenarios and their probabilities. Doing so, we find that there is a 10 per cent chance that Triangulum will collide with the Milky Way before Andromeda does, and a far greater likelihood of it joining the three-way party some time later on.

A second, quite different scenario for the Triangulum galaxy is to be ejected entirely from the local group. If it scores a glancing blow with one or both of the two larger galaxies that eventually make up Milkdromeda, it could undergo a gravitational slingshot and fly off, never to interact with another galaxy again. There is approximately a 7 per cent chance of this happening, according to gravitational models of the three galaxies.

A final possibility is an intriguing outcome. If the sun, along with the solar system, is sprayed off into space by the interaction between the Milky Way and Andromeda, perhaps we will be caught by the gravitational field of the Triangulum galaxy and 'adopted'. Estimates of the probability of the sun being adopted and becoming an orbiting star in the Triangulum galaxy are around one in 10,000, so it's a slim chance but an interesting possibility.

Whatever the role of the Triangulum galaxy, there is no doubt that the gradual creation of Milkdromeda is the main attraction. With each interaction between the two giant galaxy discs, further star formation is triggered. Gas is churned up and funnelled towards the supermassive black holes at the galaxies' centres. Giant revolving accretion discs of gas form around them, which steadily dribble material from their inner edges into the black hole's event horizon like a stream of water trickling down a plughole. The accretion discs around each black hole fill faster than they can be emptied. As gas and stars rain like a tropical deluge towards each galaxy's centre, these discs become searingly hot. Hot accreting gas splashes violently away from the core along two magnetic axes, reawakening the 'dragon' at the centre of our galaxy and reigniting the jets that formed the Fermi bubbles some ten million years ago.

As the twin cores of Milkdromeda become active, the radiation from the nuclei will instantly sterilise any planets in the jets' searchlight beams, especially in the inner regions of the galaxy. To be bathed in this radiation field will be the lethal fate of planets around a large number of the new stars, primarily situated in the inner galaxy regions, that are formed in the 'starburst' phase of the merger. Many of the far-flung solar systems (including our own) will likely be spared this doomsday glare.

After 5.5 billion years, the twin cores of Milkdromeda will most likely become steadily closer as the galaxy holds its breath for the big event—the merging of the two supermassive black holes. Sagittarius A* and its counterpart at the centre of the Andromeda galaxy (called P2) will merge to form a single black hole around ten million times the mass of the sun. As they enter a very close orbit around one another, gravitational energy will be lost in the form of gravitational waves, which are emitted in all directions at the speed of light. These waves quite literally shake the fabric of space and time, causing both lengths and times to contract and expand along the direction of the waves.

The amount by which these gravitational waves stretch and squash space across the 2.5-million-kilometre LISA detector will be less than the size of a helium atom. Such tiny changes in length are completely imperceptible on Earth, even to the sophisticated

LIGO that is doing such excellent work with shorter-wavelength gravitational waves from small binary black holes and neutron stars. We will not even notice the changes caused by these gravitational waves in our everyday lives. It will be a great time for our galaxy's physicists, however, who will be able to monitor the merger in real time using the changes in the patterns of signals from pulsars and/or changes in length measures such as lasers repeatedly bouncing off mirrors carefully placed in orbit around Earth.

What fate the rest of the galaxy?

With each successive interaction between the discs, the stars and gas become more and more disrupted. Material is pushed and pulled in many different directions by the complex gravitational interactions that occur in the creation of Milkdromeda. New stars form, and their powerful stellar winds expel gas towards the outer regions of the galaxy. As the black holes awaken, they too generate outward forces that expel gas deep into space. Our new galaxy is slowly suffocated, as with time the interstellar gas is lost and the galaxy is bloated. Stars that were shiny and new begin to age and the older stars relax into their dim, red state of retirement. The whole of the Milkdromeda galaxy loses its shape, becoming an elliptical galaxy—a bloated, spherical mass of 700 billion old, red stars.

The fate of our sun is almost certainly a large orbit of the Milkdromeda elliptical galaxy. To observers, the sky will slowly lose its contrast as the galaxy comes to resemble an overexposed photographic film. The stargazing experience of today, with a dark sky punctuated by bright stars and a narrow strip of the Milky Way, will be long gone. What has replaced it is a sky full of stars—billions upon billions of stars, distant, faint and glowing like an all-encompassing city haze. Lying on your back, you see not a single constellation in this sky, just a carpet of dots laid out as if created by the skilful pointillism of a French impressionist artist. The bright centre of our newly formed supergalaxy lights up the night, like the moon. The formation of Milkdromeda has meant the end of darkness in our night sky.

It is interesting, as we ponder the future, to consider the wider fate of our Earth, the only place in the universe known for sure to sustain life. Will there still be astronomers alive in four billion years' time, lying on the grass and gazing up at Milkdromeda? Will other facets of our night sky change on these cosmological timescales?

Let's start by considering the moon, our only natural satellite, which orbits Earth once every month at an average distance of 384,400 kilometres. Due to a subtle tidal pull between the two bodies, Earth and the moon are currently creeping 3.78 centimetres further apart per year. We can measure this by shining lasers from Earth onto the moon and bouncing their beams off special mirrors left on the lunar surface by Apollo astronauts and Soviet spacecraft. Each and every year, the time taken for the laser beams to return to Earth increases by a tiny but measurable amount. (This experiment, by the way, is further proof—in addition to the thousands of photographs and rock samples, and hundreds of hours of footage—that the moon landings in the 1960s and 1970s really happened.) The subtle expansion of the lunar orbit means that in four billion years' time its distance from Earth will have increased to 535,600 kilometres. As a consequence, the moon will appear considerably smaller and quite a bit fainter in the sky. It will also mean that the most superb astronomical phenomena that we enjoy today—total solar and lunar eclipses—will cease.

Eclipses rely on the fact that the apparent size of the sun in our sky is the same as the apparent size of the moon. There is no magical reason for this—it's just really good luck that the full moon happens to be both 400 times smaller and 400 times closer to Earth than the sun. Every now and then, the sun, the moon and Earth line up in such a way that either the sun or the moon disappears from our perspective. This doesn't happen every month because the planes of Earth's and the moon's orbits around the sun are not exactly in line with one another. When they do line up precisely, we get total solar and lunar eclipses.

To be able to see the moon completely cover the disc of the sun, showing only the beautiful outer atmosphere (called the corona), is a magnificent gift. I attended the total solar eclipse in Cornwall in the south-west corner of England in 1999, and I must say it was one of the most memorable experiences of my life. Camping in a field with a group of friends, most of whom I'd met at Space School UK, a residential experience for young astronomy nerds, only heightened the anticipation. We had set up our tents in a sloping field overlooking the English Channel on the eastern side of Cornwall, a beautiful and unspoiled county that is warmer than most of the UK but unfortunately receives its weather straight from the Atlantic Ocean. On the morning of the eclipse, clouds covered the sky and brought a damp mist to the rolling cliff-top landscape. Although we wished and willed the clouds and mist to dissipate, they did not. We watched with anticipation through special eclipse safety glasses during brief breaks in the clouds as the moon's disc slowly crept across the sun. As the time of totality came, we were dumbfounded as darkness but also a sudden coolness descended on our field and birds began to sing furiously in an eclipse-induced evening chorus.

At the moment of totality, when the entire disc of the sun disappeared from view and darkness descended on our corner of Cornwall, we stood in a circle grinning at each other like idiots and held hands and raced around to the strains of 'It's the End of the World as We Know It' by R.E.M. Around half an hour later, Sod's law kicked in and the clouds cleared to reveal a spectacular sunny afternoon. We walked to a nearby beach and made sandcastles as if nothing had happened.

Total eclipses like this are sadly a limited resource with the inexorable creep between Earth and the moon. In around 560 million years' time, the moon will be too far away to cover the sun even during a perfect alignment of orbits, and we will be treated only to annular eclipses, which happen when the moon is at its furthest point in its orbit around Earth and its disc fails to completely cover the sun. No more totality, no more dancing—it's the end of the eclipse as we know it.

Linked to the drag caused by the gravity of the Earth-Moon system, the length of the day is also increasing. Evidence from rock sediment and growth rings in coral shows that over the past 4.5 billion years, each day has become nineteen hours longer. Extrapolating that forward four billion years to the genesis of Milkdromeda, a day on Earth will be forty hours long. Scientists have predicted how such a long day might affect habitability. First, as Earth's rotation changes we could see the spinning axis of Earth tip over to a very different angle. At the moment our axis is 23.5 degrees tilted with respect to our orbit around the sun. This is the reason we have the seasons winter, summer, autumn and spring. When the axis in your hemisphere points towards the sun, you enjoy longer days and hotter weather. When your hemisphere is pointing away from the sun, you receive less of its heat and light and the temperatures are cooler.

The Earth's axis is currently quite stable: at present its axial tilt varies very slowly by about 2 degrees on a regular 41,000-year cycle. Computer models suggest that it is our tidal relationship with the moon that stabilises Earth's axis. The future increase in the distance to the moon will cause Earth's axial spin to slow down, which in turn could allow Earth's axis to tip over at an oblique angle. Any rapid shift in our axial tilt would cause sudden changes in climate to which life forms would struggle to adapt.

One final consequence of the slowing of Earth's rotation is the diminution of Earth's magnetic field. The spinning of our molten iron and nickel interior is thought to generate our magnetic field, like a dynamo on an old-fashioned bicycle light. Earth's magnetic field is significant because it acts as a protective barrier against cosmic rays and other particles from the sun that can erode our atmosphere. With the slowing of the axial rotation, the magnetic field will weaken and disappear entirely, leaving Earth-bound life forms and our atmosphere susceptible to cosmic radiation. This is yet another cosmic peril facing our planet in the future.

Human beings in their current form will not be alive to see Milkdromeda. Evolution of life on Earth happens on timescales far quicker than the three to four billion years required to change the nature of our galaxy. Even the nature of Earth will change dramatically in that time. The shifting of tectonic plates to form a supercontinent occurs on a 300- to 500-million-year cycle, bringing with it major changes to the climate, atmosphere and life on Earth. Subduction of continental plates will cause about a quarter of all ocean water to head down into our planet's interior within the next billion years. On the same timescale, the sun will increase its luminosity by 10 per cent as it runs out of hydrogen in its core and relies increasingly on helium (which releases more energy per fusion reaction). This is predicted to increase the average global temperature to approximately 47 degrees Celsius (116 degrees Fahrenheit), causing yet more water to evaporate, enhancing the greenhouse effect and leading to the complete evaporation of the oceans. By the time Milkdromeda forms in 4.8 billion years' time, the sun will be 67 per cent more luminous than today, Earth's dry surface will have melted, and the planet will have been rendered uninhabitable.

But there is some hope. Humans are an adaptable species. Between the 1780s and 1880s CE, in the space of less than 100 years, our methods of long-distance travel transformed from the horse and cart to thundering steam trains and motor cars. In the next 100 years we graduated to commercial aeroplanes jetting around the world (mid-twentieth century onwards) and a permanent human presence on the international space station (2000 CE onwards). It is not beyond reason that we could move at least a small colony of people to another planet, possibly Proxima B, or a solar system around a younger star in our vicinity. If our ingenuity allows it, our species might just as easily live in space, without a planet to call home.

Just as humans have manipulated materials to create comfortable environments, with homes heated and cooled to our whims and convenience food removed from its source, it is not beyond the realms of possibility that our descendants could create what they need from raw resources mined from space. By harnessing the enormous energy of our sun or the power of the atom via nuclear fusion, our descendants could potentially create food, water and a habitable environment for themselves without even needing a home planet. Perhaps our future stargazing rendezvous to watch the Milkdromeda spectacle will be from a viewing platform somewhere in the solar system that is close enough to the sun to harness its energy but far enough away to avoid the increasing heat and radiation it cooks up.

Maybe it doesn't matter whether it is our descendants who are viewing the metamorphosis of our galactic home. After all, there is little to indicate that Earth and the life forms that live here have any particularly special place in the cosmos. In our incredibly vast and diverse Milky Way there are an estimated 100 billion planets. To me, it seems unlikely that in the next four to five billion years none of these planets will provide the conditions necessary to spawn life. So maybe life on Earth will be a long-gone memory. Humans may have managed to leave the planet but still succumbed to the inevitable extinction that all species face.

Maybe it will be the job for another species, on another planet, in another spiral arm of the Milky Way, to gaze up at the night sky and tell their offspring a story about the glittering five-armed creature who lives in the stars. Even though we might not see it ourselves, it is the work of scientists to foretell this astonishing story about the future of life.

9

STAYING CONNECTED

In January 2018, a group of close friends and I watched a total lunar eclipse from the quiet suburban road where we live. As we sat on the warm tarmac, cups of tea in our hands, we watched the changing face of the full moon as it was overtaken by Earth's shadow. At totality, when the sun was fully obscured by our planet, the moon gradually turned orange as it bathed in the 'sunset' thrown across it by our atmosphere. It was a truly beautiful sight. Neighbours came out to join us and we all stood for a while, taking in the spectacle. We chatted about our favourite constellations and laughed and squinted through binoculars at the giant impact crater called Plato on the moon's surface. We scanned our binoculars down from the Southern Cross to peer at the ten million stars of Omega Centauri, the Milky Way's biggest globular cluster.

This is a feeling I have always loved—the kinship of friends and family bonding over the night sky—but as the world becomes more industrialised, light pollution and tall buildings are encroaching on our horizons and becoming serious barriers to casual stargazing. As well as blotting out the night sky, the widespread use of excessive and poorly designed lighting is having serious financial and environmental consequences. A recent research study on the economic impact of light pollution estimated that the USA spends nearly $7 billion every year on light that shines directly into the sky. This is a complete waste of money, and also has a serious impact on plant and animal behaviour by altering nature's reactions to the regular patterns of night and day. Using well-shielded, lower-power lighting that lights the intended target (for example, street lighting shielded to shine only towards the road) is a very simple solution to many of these problems, but sadly it has not yet been universally recognised. Legislation to control light pollution is largely absent from the statute books.

Astronomers have had a long-running battle with light pollution. The percentage of the world's people living in urban areas increased from 3 per cent in 1800 to 50 per cent in 2008, and gradually the sky has been blotted out across the globe. As cities have grown, the world's great astronomical observatories, like the Sydney Observatory and London's Royal Observatory at Greenwich (the zero longitude point on Earth, where Greenwich Mean Time originated), have been rendered museum pieces. Even modern telescopes sited in rural areas are being affected. The Australian National University's Mount Stromlo Observatory in Canberra was selected as a dark site for telescopes when it was established in 1911, but the expansion of Australia's capital city rapidly overtook the site and by the mid-1960s some of the astronomical instruments were being moved to a new, darker site at Siding Spring Observatory, nestled among the beautiful mountains of the Warrumbungle National Park.

Siding Spring is now Australia's premier optical astronomical observatory, housing the impressive 3.9-metre-diameter Anglo-Australian Telescope—the country's largest—and a host of other telescopes. But even this relatively remote site has been battling light pollution for many years. Aside from the distant glare from Sydney, increasing industrial and mining activity in the region is causing degradation in the quality of the night sky. In an attempt to address this serious issue affecting our astronomy research capability and to educate people about the negative impacts of light pollution, the site was recently awarded status as a 'Dark Sky Park' (one of only fifty worldwide) by the International Dark-Sky Association, which works to stop light pollution and protect the night skies for future generations. Despite this important symbolic gesture, many believe that the battle for a dark site suitable for world-class astronomy in Australia has been all but lost. With the encroachment of stray light and the availability of higher-altitude sites overseas, the long-term future of optical astronomy in Australia now lies primarily in the shared use of international telescopes.

When I was growing up in my small village in the north-west corner of Essex, I had a favourite stargazing spot on a private road a few hundred metres from my family home. It was not a glamorous location—it was the access track to a sewage works and a regular walking track for the village dog owners (we colloquially called it dog-shit alley)—but it was surrounded by fields, had a good clear horizon and was mine. Away from the orange glow cast by the handful of village streetlights, my eyes were free to adapt to the darkness, and the glorious contrast of the night sky unfolded readily. I learned about our eyes' reluctance to see the stars brightly, and gave my vision ten to twenty minutes to adapt fully as my pupils opened widely to allow more light to enter. I also used a red torch to light my way, which I had learned was less prone to triggering our pupils to dilate.

From my dark site I enjoyed astrophotography, another dimension of stargazing that relies on dark skies. When you take a picture of the stars, the camera shutter is usually opened for periods of up to several minutes, even a few hours if you are capturing 'star trails' (circular wheels of light from stars as the sky appears to rotate around the poles as Earth circumnavigates its axis once per day). Any stray light at all, whether from passing cars, aeroplanes or street lighting, will flood into the camera and overexpose the whole shot, ruining hours of astrophotography work.

I was lucky with my dog-shit alley observatory for a while, but when I was fourteen years old a new sports centre was built in our nearest town, Braintree, 10 kilometres from my dark sky spot. Suddenly, the southern horizon was lit up almost every evening by four glaring spotlights shining a cone of ghostly white light across miles of countryside and blotting out around 20 degrees of the sky. Because I used the sports centre regularly, it was even more annoying that my athletics training was contributing to the avalanche of light that was overtaking my astronomy site!

Angry at the loss of my beloved dark sky, I wrote to the local council about the issue. To my amazement, I was invited by the head of planning to an in-person meeting at the district council headquarters. It felt a little strange and intimidating at the age of fourteen to be in a meeting with a bunch of adults in suits, but we managed to have a good discussion about the issue of light pollution and its impacts on the environment and astronomy. I put forward some simple solutions, such as shielding that would direct the light towards the ground rather than scattering it into the air. I found examples of good lighting installed in a new housing development on the outskirts of the town and presented the development company responsible for the sensible lights with a 'Good Lighting Award' on behalf of the British Astronomical Association's Campaign for Dark Skies.

My modest local education campaign as a fourteen-year-old may have won a few hearts, but the damage was already done. Once expensive lights have been installed, owners are understandably reluctant to spend money on changing them. What is needed is best practice and regulations that provide guidance to developers and catch poor lighting in the design phase. Ultimately, people need to be given good options and financial incentives to make the right decisions. Hopefully, the peak body for astronomers, the International Astronomical Union, will continue to make progress on this issue through its program of monitoring, education and legislative protection for observatory sites. This is extremely important in order to protect our skies into the future.

A 2016 study of global light pollution revealed that more than 80 per cent of the world's population, and 99 per cent of European and US dwellers, live under light-polluted skies. The darkest skies according to this study are in inland Africa, a continent that has managed to remain largely unaffected outside a few bright conurbations such as the Nile Delta, Johannesburg and the northern and central-western coasts. Other notable dark regions are Siberia, northern Canada, the Amazon rainforest and Central Australia, which all have very low population densities. The figures are stark: 36 per cent of the world's population lives in areas whose skies are so light polluted that the Milky Way is no longer visible. In Australia the figure is 67 per cent, and in the USA and the UK it is 77 per cent.

Wherever you are in the world, despite the creeping menace of light pollution it is still possible to connect with astronomy. It's a hobby that can be done with absolutely no specialist equipment. The moon is a great place to start.

Make friends with the moon. Start by noticing its phases through the month, whenever you spot it in the sky. Is it a crescent, half, or nearly full? Can you imagine how the sun is shining on its illuminated face? Is the dark shadow across the moon to the left or to the right? Notice the time when it rises or sets. What is special about the time of the moonrise at full moon? What lunar phases can you see during the day? By mindfully experiencing the moon in this way, you can really start to appreciate our beautiful natural satellite. Try it next time you are outside.

If you do have the resources, your experience can easily be enhanced by investing in an inexpensive pair of binoculars. With binoculars or even a small telescope you can see the grey maria (Latin for 'seas', pronounced 'MAH-ray-ah') and impact craters that pepper the lunar surface. Lunar maria are not actually seas, although early astronomers believed they contained water. They are actually dry plains of basalt formed three to four billion years ago when the moon was volcanically active. Now ground down to a grey powdery dust, the dry maria form a hotchpotch of interesting features that are visible to the unaided eye (for example, 'the Man in the Moon') and look fabulous through a telescope. Perhaps the most famous of these features is Mare Tranquillitatis (the Sea of Tranquillity), where Neil Armstrong and Buzz Aldrin first stepped onto the surface of the moon in July 1969. It's possible to locate all the Apollo lunar landing sites with a lunar map and a pair of binoculars. Although you can't see any details (such as the spacecraft or leftover equipment) through binoculars or a small telescope, the Lunar Reconnaissance Orbiter has taken images of all six Apollo landing sites from lunar orbit and they are available to view online. In them, you can see the landing stages of spacecraft left on the surface as well as scientific equipment, rover tracks and even astronauts' footprints!

The planets Venus, Mars, Jupiter and Saturn can be seen without any specialist equipment. These bright wanderers through our night skies won't always be visible from your region of Earth—after all, their orbital periods around the sun range from 225 days (Venus) to 29.4 years (Saturn), so they appear to travel extremely slowly night by night through the constellations. But when they are there, you'll be treated to a bright shining 'star' for many weeks or even months on end before they move from view to another part of the sky.

Mars, with its iron-rich soil, appears in our skies as an orange-red 'star'. Through a modest-sized telescope, you can just make out the white patches at the poles made up of frozen water and carbon dioxide. You can also see darker features on the surface, which have irregular shapes that shift and change with time. In centuries gone by, these dark patches were believed to be water-containing seas or lakes on Mars. Later, with the finding that the features varied, some astronomers suspected they were caused by shifting vegetation as it grew and diminished with the seasons. Now that we have sent numerous spacecraft to Mars, we know that the dark and light patches come about during ferocious dust storms that rage across the planet's surface as dust is blown off rocky faces, exposing the darker basaltic rocks below. The dust storms themselves, which sometimes rage across the surface for weeks on end, can appear as lighter patches.

Jupiter and Saturn, the gas-giant planets, are both yellowish-white in the sky. At their visible peak they shine more brilliantly than the brightest star. Saturn's famous rings are visible with a small amateur telescope and, believe me, look absolutely amazing. My first glimpse of Saturn's rings through a 3-inch refracting telescope is still etched in my mind—mostly, I think, because they looked so thin and delicate. It was hard for my brain to process the fact that this was a natural object shaped only by the forces of physics. With a good pair of binoculars, it is quite easy to spot the largest four moons around Jupiter: Ganymede, Europa, Callisto and Io. We call them the Galilean moons because they were first spotted in 1610 when Galileo Galilei pointed his astronomical telescope to the skies. Through binoculars, they look like four tiny points of light on either side of the planet. If you look at them over a period of a few days you can see them moving in their orbits. For the best chance of spotting Jupiter's moons, my tip is to use a tripod, or lean your binoculars on a wall (or a friend's head!) to reduce the hand wobble that is magnified by them.

Venus, which is relatively close to Earth, often appears as a dazzlingly bright star close to the horizon in the morning or evening. At other times it is too close to the sun and you can't see it at all. Venus is so bright that people often mistake it for an aeroplane, or even a UFO. I love it when Venus is visible, because it has always amazed me how unnaturally bright it is considering it is only shining by the sunlight reflected from its dense atmosphere. It also reminds me that there is another Earthlike planet out there, quite close to us but with a runaway greenhouse effect that sees its daytime temperatures topping 460 degrees Celsius. We are very lucky to have access to our planet's eminently hospitable environment—too bad we seem to be ruining it so quickly. You can find information about which planets are currently visible by consulting a night-sky smartphone or tablet app, or search online for a monthly night-sky guide specific to your country or region.

Not everything in the night sky is slow and steady like the moon and planets. Periodically (several times a year), Earth's orbit will intersect a stream of dusty debris that has flaked off the surface of a comet as it orbits the sun. When this intersection happens, we are treated to a meteor shower. Meteors are tiny pieces of rocky debris that streak through our sky. One minute they're floating around our solar system, the next they find themselves hurtling through Earth's atmosphere at 50 kilometres per second. As they rub against the air, friction heats the rocky material to a temperature of 1500 degrees Celsius. The rock quickly erodes and breaks up, burning and barrelling through the sky and leaving behind a red-hot trail. As an observer of this phenomenon, you see a brief yet quite bright streak of light that is commonly called a shooting star.

Meteors can appear completely at random, at any time. They don't land (those bigger rocks that make it all the way through the atmosphere are called meteorites) but burn up completely in the sky. During a meteor shower, many meteors appear over a period of two to three nights from a particular direction in the sky. Each shower is named after the constellation from which the meteors appear to radiate. During a Geminid meteor shower, for example, meteors radiate from the constellation of Gemini, since that is the position where Earth's atmosphere intersects the comets' path.

Occasionally, every few decades or so, there will be a spectacular 'meteor storm' with shooting stars raining down every few seconds. The last great meteor storm was in 1966, when the Leonid meteor shower put on an unbelievable display and shooting stars in their thousands fell from the sky. Some described it as looking like a waterfall, as the display peaked at around forty meteors per second. Displays like this are rare and quite difficult to predict, although they are thought to be linked to the closest passage of a comet to the sun. I am pretty devastated that I've never seen a meteor storm despite staring at the night sky for many years. It just goes to show that it's one of those astonishing features of the natural world that you can't control and are incredibly lucky to witness. The next Leonid storm is predicted to occur in 2034, and I hope to be around to see it. If you don't want to wait that long, you can look up the dates of annual meteor showers on the internet and enjoy the spectacle.

An excellent way to get connected with the stars is to hook up with your local astronomical society. There are thousands of amateur astronomy groups around the world, full of people who are committed to sharing their passion. Most have monthly meetings featuring talks by local experts or, sometimes, professional astronomers. They are great places to discuss various astronomical topics, and many have specialist equipment that they are happy to lend or share. All my early telescope observations were enabled by my local astronomical society, who lent me their 10-inch reflecting telescope for a period of four weeks. It was a white tube as long as me with a large mirror to collect the light and a smorgasbord of different eyepieces to achieve different levels of magnification. It was a wonderful resource for me as I hungrily learned to navigate my way through star clusters, planets and nebulae. At night I followed the changing positions of the moons of Jupiter and sketched the Orion nebula and the trapezium star cluster at its centre, and by day I followed sunspots across the solar disc by projecting the solar image onto a piece of cardboard. My adventures with this large telescope gave me practice in locating sky positions and tracking objects as they moved throughout the night. I also gained a good understanding of how telescopes work, including optics, their mountings, which eyepieces to use, and how to photograph various objects in the sky. All of this gave me a deeper level of understanding that helped me in my professional career, even though all our processes with modern telescopes are now computerised.

The night sky has been an important part of human civilisation for as long as people have gathered and communicated their stories. Using science, we have progressed from spinning tales about the stars to building telescopes that can help us unravel the remarkable truth of the stars' existence. With science comes technology, some of which has been used wastefully and to excess. The unintelligent use of lighting has led us to unwittingly blot out the stars. Will the future of stargazing involve jumping on a commercial spaceflight and staying in an Earth-orbiting astronomy hotel with a dark-sky viewing platform where we can bathe in the wonder of the stars? Let's hope not. After all, it is possible to undo the damage we have done. If human populations can find more efficient and intelligent ways of deploying the new street lighting that is required in populated areas and replacing old, inefficient lights with more suitable versions, we will once again stand a chance of preserving dark rural areas in which people can bathe in beautiful, natural darkness. Otherwise, we are destined to light up our world for centuries to come and 99 per cent of the world's population will not be able to look up and notice the incredible universe in which they live.

Our night sky is precious. For most people on our crowded and light-polluted Earth, opportunities to ponder the meaning of existence while gazing at a dark night sky are severely limited. Not only is our atmosphere getting lighter, but the entire galaxy is too. As the Milky Way meanders towards its inevitable collision with its cosmic neighbour, the days of dark skies filled with pinpoints of light tracing out constellations are numbered, wherever you happen to live within our galaxy. The whole character of the sky will go through a phenomenal transformation. On approach, Andromeda will begin to dominate the sky as a gigantic swirling spiral appearing as a co-star beside the Milky Way. As the collision induces a starburst, bright blue stars will rapidly burst forth like popcorn and the constellations will change dramatically. As our galaxies bend and break together, the limbs of the Milky Way and Andromeda will stretch from horizon to horizon, blotting out much of the background and reducing our visibility of the distant universe like a winter fog. Eventually, our whole sky will be enveloped by the uniform glow of trillions of stars as Milkdromeda settles down for its long and glittering retirement.

However you look at it, now is a golden age for astronomy. We exist here on Earth as the only known astronomers in the universe. Our position in the Milky Way and the epoch in which we live mean that the sky is neither empty and black, nor filled with all-consuming starlight fog. Our skies afford us a view of hundreds of millions of distant galaxies, from which we have been able to construct a theory of our universe's origin and evolution. If we existed much later in history, we may not even know that there was ever anything else in our universe outside of the towering dome of blackness that we call the night sky.

Our epoch on Earth is auspicious, too. We live around a stable star in the middle of its lifetime that provides the perfect conditions for our existence. By a bewildering coincidence, the moon lines up regularly and perfectly with the sun, providing mind-blowing eclipses for us to enjoy. Meteors rain down on our planet with regularity, providing a safe yet intimate congress with the four-billion-year-old dust from which our solar system was born. Planets dance through our skies, twirling around the sun with pace and tempo that reveal their positions in the cosmic order. Galaxies poke through the cloak of invisibility and reveal hints of the vast scale of our cosmos. Humans still walk among us who have broken the bonds of Earth's gravity and have stood on the face of the moon.

We are all children of the universe, and it's time to meet the family. Next time the sky is clear, get a star map (or an app), grab someone you love and take a walk under the stars. Go out into the backyard and get acquainted with your celestial neighbours. Marvel at the sparkling planet Venus. Consciously observe the changing of the moon's phases. Take a drive into the country with a flask of coffee and a blanket and treat yourself to the dramatic spectacle of the universe. Sit and wonder and imagine like your ancestors did. Because for your descendants in a few billion years, this will all have been blown away.

ACKNOWLEDGEMENTS

The author would like to thank Sally Heath, Louise Stirling, Katie Purvis, Emma Rusher and everyone at Melbourne University Publishing for making the creation of this book a delightful experience.

To everyone who has ever stoked my fire for astronomy, from Patrick Moore, to Braintree Astronomical Society, especially Josh Greenwold and Space School UK, you did a great thing. Special thanks to my parents, you gave me so much of your time and attention. To Hannah Worters, who is my bestest space buddy to this day, thanks too for sharing my hobby all the way from the tower blocks of Osterley to the early morning start of the Comrades Marathon in Durban (look! There's Vega!).

Biggest love and thanks go to my partner Michelle Reid, who is a wonderful sounding board for my ideas and who is my lifelong companion for watching the stars. And to Inky, who still manages to inspire us every day.

INDEX